earth user's guide to *teaching permaculture*

Rosemary Morrow

Permanent Publications

Published by:
Permanent Publications
Hyden House Limited
The Sustainability Centre
East Meon
Hampshire
GU32 1HR
United Kingdom
Tel: 01730 823 311
Fax: 01730 823 322
Overseas: (international code +44 - 1730)
info@permanentpublications.co.uk
www.permanentpublications.co.uk

Distributed in the USA by
Chelsea Green Publishing Company
PO Box 428, White River Junction, VT 05001
www.chelseagreen.com

First published 1997, fully revised and updated 2014

Designed by
Two Plus George Limited
www.TwoPlusGeorge.co.uk

Cover design by
John Adams, drawings by Gabrielle Paananen

Illustrations by
Gabrielle Paananen

Index by
Amit Prasad, 009amit@gmail.com

Printed and bound in the UK by
Cambrian Printers, Aberystwyth

Printed on paper from mixed sources certified by the
Forest Stewardship Council

The Forest Stewardship Council (FSC) is a non-profit international organisation established to promote the responsible management of the world's forests. Products carrying the FSC label are independently certified to assure consumers that they come from forests that are managed to meet the social, economic and ecological needs of present and future generations.

British Library Cataloguing-in-Publication Data
A catalogue record for this book is available from the British Library

ISBN 978 1 85623 145 9

Contents

Dedication

I dedicate this book to my committed, funny and passionate nephew, Michael, who made all our lives better and lived permaculture. Michael had a profound love of all living and non-living Earth. His miracle was Life.

SECTION 1
Introductions and starting well

Letter to teachers

Dear permaculture teacher ...

In the 17 years since I wrote the first edition of this book the world has changed inconceivably, and the speed and direction of change is alarming. However the permaculture curriculum is still excellent and needs little alteration.

What students require from you

The major outcome for students is their ability to evaluate landscape and redesign it for a robust and resilient future. You need to get them to this point. Your students will mirror your knowledge, attitudes and skills when they finish their classes.

Teaching permaculture requires you to have:

- Sound unbiased knowledge, clear teaching goals, strategies and methods
- Clear understanding of the skills and attitudes students will be competent in by the end of your course

If you don't have clear teaching goals and student outcomes you won't be able to monitor their progress and your students won't be competent permaculturists.

Start by telling students that they won't be spending much time getting their hands dirty. These 'bootcamp' skills follow the Permaculture Design Course (PDC) and will occupy them for the rest of their lives. Instead, in your course, they will learn to see, consider, analyse and design productive landscapes from balconies to large farms and towns. Permaculture uses low technology and high science, and the PDC always remains solidly based on these tangibles.

Resources and opportunities for teachers

There is often little opportunity for professional exchanges among teachers on their methods of teaching, their presentation of material and quality control of knowledge and skills. I speak of the 'loneliness of the Permaculture Design Teacher' because many of us seldom have opportunities to see how we can improve or change our teaching to be more effective. We do not always know if we are teaching to a high standard, and how much we are 'in sync' with other teachers. We all want to teach well to a good standard. Some of us have no teaching experience; others have some experience in teaching but have not been very long in permaculture. Most are strong in some areas yet weak in others. Some would like to teach but have not dared until now because it seemed overwhelming. I suggest that we teachers offer places for trainee teachers to learn.

This book will assist you to put together a set of teaching notes. After teaching permaculture for nearly ten years it took me another two years to assemble and order this information and it will never be finished.

You will want to develop your own notes to suit the courses that you run. The basic syllabus does not need to change. It is gift to us as teachers and to the world. My teaching style and methods change appropriately for different topics, as will yours.

Course notes have been produced by Robin Frances in Lismore, New South Wales, by Sky formerly of Crystal Waters Queensland, and by teachers in other parts of the world. I encourage you to look at them. Robin Clayfield offers courses in creative facilitation relevant to permaculture. Other teachers have put their curriculum online.

Can the PDC change?

Although the ethics and principles of the PDC remain unchanged and enduring, the structure and flow can change. I prefer to develop landscape reading and analysis skills first as design begins immediately with analysis, such as in map reading and sector analysis.

The content and teaching methods, strategies and techniques in permaculture have been corroborated as new information about the environment surfaces. The ethics and principles of permaculture stay constant, and now much of the content given in early courses has been corroborated and extended by recent research and practice. For example, zone III now has principles and examples beyond Fukuoka and alley cropping, including pasture cropping, contour forests and provides solid evidence for cell grazing and pasture cropping which was not available for the first edition of this book. New cities and city regeneration now have fine principles to assist permaculture designers.

Who can teach?

By permaculture tradition anyone who has a PDC can teach permaculture and use the word permaculture which is invested in PDC graduates. Some institutes and organisations now desire diplomas and for graduates to wait two years before teaching. I think it is better to start as quickly as possible to maintain enthusiasm, knowledge and motivation. The two year rule simply doesn't make sense for some people who arrive extraordinarily well equipped to start teaching, and for others where the national need and demands is very great. They need another pathway.

The adult learned centred approach

The difference between this book and other teachers notes is my emphasis on adult learners. I call this learner care based on Care of People and it entails teaching permaculture based on the second permaculture ethic. Being passionate about teaching adults, I use the most recent research on how adults learn and I find this approach not only ethical, but also extremely effective.

We don't teach religion or psychology, only Care of People as best exemplified by our behaviour towards students and other: Treat them well and require them to treat each other well.

The structure of these teaching notes

The choice is yours. You can work your way from the beginning to the end and you can follow my timetable (see Appendix). Or, you can draw up your own timetable and then select the units to give continuity and depth to your teaching.

The sequence you find in these notes is a result of teaching qualifications in Adult Education and especially in the field of Non-Formal Education (NFE). It consists of sections assembled to give continuity, breadth and depth. The sections are:

Sections	Title	Units
1	Introductions and starting well	Two essays: Letter to teachers Profile of adult learners
2	Beginning permaculture	1-5
3	Ecological themes	6-13
4	Designing productive landscapes	14-22
5	Increasing resilience	23-30
6	Social permaculture	31-39
	Last PDC day	40

I have found it easier, although perhaps less exciting for some students, to start by giving the whole class the same grounding in the applied sciences of the Principles, Ecology, Soils, Water, Energy, Plants and Forests. Then I develop the design sections of the course when the concepts, themes and principles are well established.

The importance of permaculture teachers' ethics and principles

'Central to permaculture are the three ethics: earth care, people care and fair shares. They form the foundation for permaculture design and are also found in most traditional societies. Ethics are culturally evolved mechanisms that regulate self-interest, giving us a better understanding of good and bad outcomes. The greater the power of humans, the more critical ethics become for long-term cultural and biological survival. Permaculture ethics are distilled from research into community ethics, learning from cultures that have existed in relative balance with their environment for much longer than more recent civilisations. This does not mean that we should ignore the great teachings of modern times, but in the transition to a sustainable future, we need to consider values and concepts outside the current social norm.'

David Holmgren
on permaculture ethics

We teachers endeavour to live the three permaculture ethics. They become particularly important when we are teaching permaculture. We know that our students imbibe our behaviour and, consciously or unconsciously, copy us. We are their models.
So I have tried to articulate teaching ethics for my goals and behaviour. You can write your own. It is a searching exercise and I show them to my students so I am accountable to them for what they learn and how they are treated. They are further broken down into principles.

I have a set of draft teaching objectives for the PDC.

Teachers' ethics

- Facilitate ethical conduct and equitable relationships – See Class agreements;
- Foster commitment to meaningful collaboration and reciprocal responsibility of all participants;
- Deliver clear complete, accurate permaculture content and process with attention to the three ethics

Teacher's principles for learning

Encourage, model and create a non-threatening learning environment through

- Active participation
- Confidentiality and diligence
- Respect
- Trust and protection

- reciprocity and acknowledgement of contributions and due credit.

Always demonstrate continual willingness to evaluate your own understanding, actions, responsibilities and knowledge.

Teachers' principles for earthcare content

- Reinforce local culture
- Give only relevant and accurate information
- Acknowledge and recognize prior suffering/injustices
- Apply environmentally and socially sound principles
- Encourage information to be locally adapted
- Build on existing valuable knowledge and practices

Environmental goals – Care of Earth

- Autonomy in food, water and energy through sustainable methods
- Small scale and long-term productive designs
- Close the hunger gaps and periods of insufficiency
- Rejuvenate landscapes

Social goals – Care of People

- Encourage active co-operation
- Empower people to act
- Acquisition of useful peaceful skills
- Develop and extend local knowledge networks and trust

The units can be used many ways. For example, you can select some to teach an Introductory Course.

Each unit provides the essential skeletal information required for teachers and students. Each unit consists of five to eight pages and within each there is a format which consists of the following:

This unit situates the importance and value in the whole course and how it links with other units.

Learning objectives given for key areas in the topic in which students need to be competent. They are reflected and reinforced in classroom activities and student activities.

Graphics Students learn more from pictures than words. Here are ideas for photos, slides, overhead projection sheets, drawings, wikipedia graphs you could use to improve learning.

Find line drawings to copy onto boards, to overhead projectors or to photocopy for your students according to your teaching method. Only a few drawings are provided since you are encouraged to routinely research and develop your own.

Content summarizes the main points students should know including, Ethics, Principles, and factual information.

Student activities for students to do at home to reinforce their class learning and to acquire extra skills. These support the learning objectives.

Resources are the videos, places, people, and TV programmes and websites. The list indicates a few possibilities, however you need to build your own resource list – often specific to your bioregion. You always need to know more than you teach.

References consist of booklists, with the principal books for the unit listed first and then the extra reading. You should frequently up-date your knowledge. Advances and additions to knowledge in areas such as renewable energies, LETS and financial systems, and recycling need to be kept up to date.

Teaching methods and tools

If you are not experienced or have not thought how adults learn differently from children, first read the article on Adult Learners as a background. Some highly trained teachers are not good teachers and some people who are not formally trained are very gifted.

The main principle in teaching methods is that you vary your methods during a session to reach the different learning styles of your students. For example:

- Group Discussion
- Lecture (always short)
- Brainstorm
- Seminar
- Field visit
  Demonstration
- Questioning
- Your ideas

The critical factor is that a variety of approaches is required to meet the needs of the people who learn in different ways (different learning styles). Teaching methods, to be effective, must be learner-centred, not information delivery centred. When students are actively participating you can be reasonably confident that they are learning. It is more important that students are able to learn and make sense of each unit (process and internalize) than you tell them everything that you know.

This guide to teaching permaculture has been designed as a handbook for the full 72 hour Permaculture course which gives students the international qualification presently known as Permaculture Consultant's Design Course certificate.

This book is not prescriptive. Alter it for different teaching situations, cultures and biozones which require distinctive approaches and techniques. Open class debates and give more bite to weight, depth and relevance of each topic.

Course participation

Courses are open to anyone who wishes to attend, without discrimination of age, sex, religion or education. Teachers set their own fees for their courses. However, recently course fees have been too high for marginalised or disadvantaged people who often have most need of permaculture. It is always a balance between making a living as a teacher and charging fees people can afford. Traditionally, one free or reduced fee place is given to a student in each course.

Course outcomes

The table below shows the outcome of what to expect from your graduates. I work to these outcomes and constantly monitor students.

To demonstrate the fundamental permaculture design skills, practices and outcomes required by a competent PDC graduate, I require two assignments.

1. A home design which hones analysis and demonstrates a design ready for implementation
2. A group design which develops diverse group skills and professional activities

Then:

- Plan transition from what is present to a concept plan for 15 years in the future.
- Order of work plan – site analysis, design and implementation.
- Design: Scale/ pens/ pencils/ shading/cross-sections/colour/ pictures/ key/ materials/ blow-ups/ transepts

Practical work and skills acquired through sitting a PDC

- Sector analysis
- Microclimate analysis
- Whole site water plan: source to sink
- Whole site soil analysis and zone nutrient planning
- Whole site plant identification, propagation and design
- Sheet mulch garden in Zone I
- Site analysis and whole site design
- Weeds, IPM, disaster and wildlife analysis and restoration
- Detailed design of zones and whole site to concept standard
- Inventory of bioregional economies, tenure and invisible structures
- Analysis of urban, village and neighbourhoods and transition to restoration for sustainability and resilience

Themes	**Analysis of**	**Design to**
Sectors	Wind, fire, sun, water, all forms of pollution, wildlife, pests, etc.	Deflect or Use
Maps	Contours, maps, aspects, slope	Use useful land, or reserve it for protection to filter or modify elements.
Water	Full audit domestic and rural	Restore and clean water in topography rivers, lakes, keylines, aquifers
Soil	Zones I to V, soils audit	Site nutrition via zones
Climate	Sun, wind, radiation and their impacts	Windbreaks, structure, plant placement, risks
Microclimates	Vegetation soils, water, structures, topography	Thermal belts, cold sinks, terraces/swales, avoid or enhance microclimates
Disaster	Risk and analysis	Avoid or endure two most likely disasters
Produce a concept design – use graphics, models, drawings	Site analysis through models, drawings and demonstrating permaculture principles	Concept design derived closely from the site analysis
Strengthening design	Integrated pest management, aquaculture wildlife	Include these elements to support the design
Social	Economics, city, village, neighbour-hood design	Bioregions, ethical income and money, community economics, land tenure

Site visits

Locate and visit permaculture sites to see different approaches to sustainable house and land design, and water/energy and food systems.

We all dream of a restored, healthy world. For me, a permaculture teacher is one who dreams their dreams and for many people today, permaculture is their dream.

I offer these two thoughts:

'You can waken people only by dreaming their dreams more clearly than they can dream them themselves.' Russian, Alexander Herzen

'The greatest gift you can offer is that of useful knowledge.' Fritz Schumacher

I wish you joy, inspiration and great satisfaction in your teaching.

ROSEMARY MORROW
April 2013

A profile of adult learners

Think of your PDC course as a learning ecosystem; a social ecosystem. It has many elements and they are interdependent and inter-related. i.e. teaching venue, students, teachers, teaching methods and teaching aids. We explore each of the elements. Each will impact on the others and on how effectively students learn. Remember teaching doesn't automatically ensure learning. There are more elements than the teacher alone.

We know some things about adult learners. We know why they come to courses, what their expectations are, and to some extent, how they learn.

Adult learners are different from child students and when you understand these differences you improve your effectiveness.

My main objective as a permaculture teacher is to assist adults in learning what they want and need to learn. To reach the different adult learning styles you need different methods and different teaching aids. And, I frequently remind myself - I do not have to tell them all that I know. Often they know more than I do.

Start by building a learning ecosystem and a code which enables everyone to participate and learn. This is the one I use.

Learning agreements to practice Care of People

These agreements originated in the peace movement and are used in this form or one close to it, in the Alternatives to Violence Project (AVP) and also by mediators and negotiators. They induce respect for each other and also encourage participants to practise the permaculture ethic, Care of People. They are very effective and can be re-introduced at any time when classes are going off-track. Learners will usually self-correct their behaviour. Of course the facilitator/ teacher is the main model closely watched by the class to see how the teacher lives them. Respect is usually demonstrated as good listening, giving attention and observing one's own contribution.

We are all teachers and learners

- Everyone has a right to speak
- No put downs, ridicule, of yourself or others
- Respect the airspace – don't talk over each other and don't speak too long. Stick with the topic

- We are all at different places on the journey
- Co-operate not compete
- Secrets stay here – confidentiality
- Speak only for yourself
- Everyone has the right to pass

Nearly all your students will be adults, that is, over 18 years old. Adults, like teachers, have considerable responsibilities. They can have heavy commitments. Many are returning to education and feel nervous. Some are working full- or part-time. Others have children and partners. They may not have much time. Some have many resources and others, almost none. Their strength is that they have volunteered to come which means their initial motivation is high. They like to have some control over classroom and decisions. Adults need to succeed – as Carolyn Nuttall pointed out to me.

As adults they usually prefer:

- To be self-pacing in their activities
- Individualised responses from you
- Student-centred learning because they value autonomy and responsibility in their learning

Each student will want different outcomes from the course.

Facilitation needs to supply pleasure and to support motivation. If you depersonalise interactions with participants they may feel snubbed or not worthwhile. The best teachers:

- Adopt a conversational style
- Know something about the student's strengths and weaknesses
- Can direct them to their special interest

Feedback is a necessary part of a two-way communication that urges students to search out and use relevant material. Feedback which asks deeper questions encourages greater originality.

Destructive criticism, sarcasm, showing off and insensitive remarks often lead to students dropping out of courses. If lessons are dull, sound like a textbook, or use language not easily accessible to students the chances of good learning are severely diminished. Using concepts and language they have not learned is just rude. Endeavour to:

- Make praise personal
- Show your own personality
- Practise what you preach
- Use supportive comments not judgemental ones

Achievement is the most powerful motivator

Don't be afraid to praise: Praise initiative, independence, critical thought and willingness to learn – perseverance. Pointing out errors has no positive effect whatsoever. Adults returning to study after a long absence generally lack confidence. They require reassurance. A kindly remark or a genuinely helpful, reassuring comment may be all that is needed to sustain their enthusiasm.

Learner-centred teaching:

- Removes the obstacles that go with traditional classrooms such as bells and nasty statements
- Allows learners to control to some degree what they learn, and how they learn it

However, the PDC also licenses design graduates and so you must ensure they leave with clearly defined knowledge and skills – and morally, this is only a fair return for their fees and attention to you.

So, to suumerize, research indicates that adult learners:

- Like to be in control of their own learning
- Don't like to be ordered to do something
- Like reasons and where possible, alternatives
- Don't like to be patronised, or treated as an impersonal statistic
- Prefer the informal, friendly, personal approach
- Dislike criticism even when well-meant.

Don't ever bother to show off your knowledge to adult learners. They will rarely be impressed and are more likely to be intimidated or infuriated.

PDCs for special purposes are courses are increasingly offered to a wider range of learners eg. people who are:

- Isolated geographically
- In institutions, e.g. Prisons, hospitals
- Shift workers
- Have disabilities
- Unable to access childcare
- Uncomfortable with formal classes
- Learning through an interpreter
- Illiterate
- Traumatized by war or disaster.

In these cases it is most important that you:

- Make your comments positive and effective
- Adapt strategies which benefit your type of students
- Comment in a simple, personal and conversational style – use student's names
- Encourage positive attitudes to learning
- Offer appropriate support
- Take every opportunity to draw upon the immediate environment for illustration
- Present difficult concepts in more than one way
- Don't correct grammar or spelling
- Know your students' special interests.

Your classes need peer expertise and peer support for each other and often just plain fun, with considerable personal interaction. This greatly assists learning, although is apparently unrelated to the formal process of teaching.

Other issues which assist teachers and learners:

- Teachers' voice and intonation
- Non-verbal communication in body language and facial expression
- Peer competition
- Outside knowledge and experience
- Good physical environment, i.e. light, toilet, tea etc

Time management

Special students:

- Make allowances but don't talk down, comments should be appropriate
- Have commonsense and imagination
- Disability – take care with practical work
- Indigenous peoples – be sensitive to background – use co-operation
- Female students – sometimes have less confidence and self esteem

Teacher's fundamental approach:

- Know the learning material very well
- Motivate, encourage and praise students
- Avoid criticism, be friendly and personal
- Retain student interest in the course
- Be constructive
- Refer students to other sources
- Encourage further learning
- Answer queries promptly and appropriately
- Advise students on their hopes
- Advise students on their progress, special abilities and skills
- Set objectives and review them
- Appropriate language and vocabulary
- Clarity of explanation
- Specific references
- Positive reinforcement of individuals
- Friendly supportive style

Are you challenged? And you still want to teach? Well, go for it. You'll love it and is anything better than permaculture?

Acknowledgements

The Vietnamese say, "You cannot do anything alone". I find you cannot write a book alone. So many people of goodwill and concern have helped me with advice, support, time and ideas, so this book really is a collective effort. I love to talk about permaculture, its idea, and how to teach it with passion and unending interest, so along the way I have been grateful for conversations with Carolyn Nuttall, Lis Bastian, Fiona Campbell, Robin Clayfield and others who are intrigued by how people learn. We still know almost nothing about how people learn and what makes one active and the other sit down and think but we know when it happens and we are elated!

Mostly, it has been my students who have challenged me, tested my theories and broken them, joined in enthusiastically and gone home and done brilliant work with panache and creativity. I welcome those who challenge me. There are the new teachers such as Alexia Gratelle, Paula Paananen and Alfred Dekker and now younger ones such as Leo and Philippe. They surpass me and filled me with joy.

I know we teachers of permaculture are on the trail of something profound when we practise the second ethic, Care of People, in our teaching and although it appears easy, it is extremely testing.

Finally, all errors and omissions are mine.

And so simply,
Thank you all so much

About the Illustrator

Illustrator Gabrielle (Gabby) Paananen is a 15 year old artist and mycologist who practices permaculture at home and in communities in Africa. She is passionate about engaging her generation to care about the earth and each other.

SECTION 2

Introduction to permaculture

UNIT 1
Your first class – meeting each other

In this first unit you and your students meet and get to know each other. You will gain an idea of the range of experience in the class and what each person hopes to gain from the course. Your objective is to build confidence and trust among the participants which will be effective throughout the rest of the course.

In this first unit you cover:

- Introductions through various activities
- Course information and requirements
- Global threats – if you have time or cover it in the next unit

Discuss housekeeping issues such as the course content, timetable, assignments and attendance requirements. With learner centred facilitation, laptops and computers are not appropriate in the classroom because we want maximum eye contact and interaction. People with computers look at screens. You could give some time for this before breaks or at the end of each day.

Learning objectives

By the end of this unit students will have:

- Met each other
- Stated what experience and hopes they are bringing to your class
- Listed the main global and local issues facing them
- Clarified the scope of permaculture in resolving these issues

Teaching tools

- Have a chalkboard, white board, even brick wall for students to list their names and interests
- List words associated with environmental destruction
- Sketches of greenhouse effect or global warming
- For global problems I recommend the DVD Tide of Change. Show for 11 minutes then discuss

Ethics

Begin building a co-operative, non-threatening and trusting learning environment. At this stage initiate and practise behaviour which demonstrates Care of People.

Principles

- Every adult student brings knowledge and many experiences that the class can share
- Every person should speak at least once in each class (otherwise you end up with only talkers while the silent ones are overlooked)
- The larger issues of climate change, land degradation and decreasing biodiversity are everyone's problems with local and global impact

The PDC is not a boot camp of techniques. It provides principles for approaching global and local problems. Students will learn to see, interact, analyse and repair human and animal living systems.

Terms

When participants don't understand words and concepts they switch off to them. When you revise words it helps to ask learners to explain them before you use them further to you to make sure they have understood what you are talking about. Revising vocabulary introduces concepts they will need.

Agri-business The economic and industrial empire that runs monocultures. Its proponents see food and the environment not as resources but ascommodities. It operates on fossil fuels.

Chemical free Not requiring artificial, industrial chemicals such as fertilisers and biocides that use large quantities of fossil fuels.

Diversity Of a species, variety, ecosystem, work, housing; having many forms.

Monoculture A single crop or enterprise, highly vulnerable to failure, destruction and energy consuming.

Self-sufficiency A system requiring inputs to totally maintain itself. The biosphere and its ecosystems are self-sufficient with inputs of solar energy, rain, seed, wind and other factors. This is very difficult for humans to achieve.

Stable Able to perpetuate itself without loss of species or requiring artificial external inputs; in equilibrium.

Sustainability When soil, water, biodiversity, animals species and other resources are never diminished. All now need replenishing.

Ask the class for definition of sustainability and compare it with self-sufficiency.

Activities for introductions

Introduce yourself and describe your background while modeling a short clear introduction. Ask them to introduce themselves in one of several ways:

- *Their hopes for the course*
- *Any relevant background*
- *Learners then introduce what they think permaculture is*

1. *Learners introduce themselves to their neighbour and the neighbour later introduces her to the group.*
2. *Each one speaks to the group when they feel confident, or,*
3. *Each student stands and has ten seconds to introduce themselves. They do this until everyone has introduced themselves to everyone else.*
4. *Write their name and three things about themselves on a piece of paper and then hold it up and speak to it.*
5. *Any other introductory activities you know which have people interacting and lets everyone be seen and heard.*

Each student writes their information on the board so every student can read it. This also helps you remember names and later direct special information to students. Learners can make a note of others with similar interests or special knowledge and experience and chat to each other for a few minutes after the activity.

Name	Hopes	Experience
Sally	Grow own veggies	Home gardening
Tan	5 acres, someday	Community gardening
Joss	Self-sufficiency	Horticulture
Maria	Live in community	Farmer
Abijit	Clean food	Grows own veggies

This gives you an idea of the range of experience in the class and what each person hopes to do with the knowledge gained from the course, e.g. to teach, work in the tropics, practise commercial organic food growing, become a permaculture designer...

Course requirement information

If you have a host for your course they can do the housekeeping which covers such items as:

- Phone use
- Help with meals
- Private areas
- Washrooms/toilets
- Room cleaning
- Recycling
- Zero waste
- Security
- Meals

Learners feel secure when told what the requirements are for a certificate.

You must outline the Certificate requirements covering:

- Assignments – 80% References
- Punctuality
- Excursions
- Note taking
- Hand-outs (I give only two handouts)
- Timetable – stuck up in room
- Attendance
- Practical and theory work
- Final party/celebration

Global threats activities

This topic is often called 'Permaculture Opportunities'* or Global Problems and needs to be covered very early in the course. Sometimes learners feel depressed. I explain that this is the low point in the course and that it will be positive after this unit with all permaculture concepts and activities being directed at solutions.

> I think that I shall never see,
> A billboard lovely as a tree,
> Perhaps unless the billboards fall,
> I'll never see a tree at all.
>
> Ogden Nash

1. *Write the following list for the class and give a few minutes for discussion. Write students' correct definitions and add extra information if required.*

   Ask which of the issues are 'local' and what specific effects they have. Ask students to connect their everyday lifestyle to global issues. Discuss tipping points.
   - *Fresh water depletion*
   - *Land degradation*

* Thanks to Tamara Griffith who gave me this way of thinking about threats to global survival. I like it very much. It exemplifies a permaculture principle.

- *Loss of genetic diversity*
- *Global warming*
- *Growth and development economics*
- *Peak oil, food, population, soils*

2. *Using the same topics, ask people to divide themselves into groups with each one taking a different topic. Then they will look at the causes and consequences of their chosen topic. Each group reports back to the class with each person in the group reporting one item or thought from their discussions.*
3. *Give a case study such as that on Viet Nam (see at end of unit) and ask students to discuss it according to the topics above.*

Importance of the ecological footprint

By this time some students are feeling despondent about the future of humans and life as we know it on Earth. It may be worthwhile to spend some time going through a few options for staying positive and active, such as religious faith, power of one, the mercy of nature, and the evolution to a less consuming species, effectively reducing our load on Earth. This discussion is valuable for the whole class. Use researched images for illustrations.

Ask students to measure their eco-footprint. Look up different types and evaluate them.

Draw a big footprint on the board.

The ecological footprint as a measure of an individual, city, organisations, or clan's consumption is expressed in hectares. This is a calculation based on the five 'toes' e.g. Consumption of energy, water, food, transport and housing. Ideally every person on Earth would use the resources of 1.8ha to meet all their needs. This would be sustainable. Australian footprint is about 7.8 ha – which requires three Earths if they continue to devour resources at their present rate. A Bangladeshi child's footprint is about 0.2 ha. For us with the big footprints, its value is that:

- We can know what our total consumption is
- We can see where we consume the most
- We can then regulate this and make a difference
- We can reduce our footprint

The permaculture course is concerned with reducing our footprint and actively replenishing resources.

The ecological footprint has emerged as the world's premier measure of humanity's demand on nature. The process of measuring our own ecological footprint helps us to understand how our everyday choices and activities contribute to our ecological impact. Ecological footprint reports are produced annually by The Living Planet and give consumption in global hectares per person.

Understanding the scope of permaculture

Many learners come to your course in trust, knowing little about permaculture. So to elicit their understanding you can do an activity:

Use a board/paper with a circle with the word permaculture in the middle. Now ask students to write words they associate with permaculture as a brainstorm. Encourage more words and maximum individual participation. At the end, ask learners to go back and link some of the words.

Your role is to explain why some ideas are not permaculture e.g. astrology and water diving. Explain that permaculture is about 'tangibles'. You may need to add to what the learners have written.

Permaculture is not only a gardening or a farming method. It is much more comprehensive and impinges on all issues and can ameliorate them. Permaculture is also about community design, ethical investment, invisible structures, urban and rural planning etc. It is imperative that learners begin their observations of the environment in every facet. Permaculture is deductive not prescriptive and so requires good analytic and observational skills. From observation, students can develop theories and gain solid empirical data. Discuss how chemical applications, insect treatments etc are all prescriptive and, by contrast, Permaculture requires thoughtful and accurate observation and deduction (use examples from Fukuoka).

Finally... Thank the class for their time, attention and contributions and read them a passage from Chief Seattle, Julienne of Norwich, or the astronauts on reaching the moon e.g. 'This Earth is all we have.'

Student activities

Optional student activities consolidate effective learning. They are highly recommended.

Learners can try these:

- *Stare out of their windows at home and make a list of all the things they see, feel, taste, smell and hear. They should walk outside everyday, observe and integrate their observations*
- *Keep an environmental diary which records their observations*
- *Keep media clippings or a notebook of environmental issues and classify them e.g. atmospheric, soil degradation*
- *Try to correlate observations e.g. hot dry days with an increase in red spider mite*

Notes for teachers

Some students will change their behaviour and outlook. One of your aims is to assist learners change from being prescribers and consumers to observers, deducers and conservers, with a corresponding increase in their quality of life. Students who start by speaking a great deal may become silent and silent people may talk much later in the course. Wait and see what happens. Don't make decisions about people on the first day or so.

In conclusion, a permaculture course can be overwhelming to some learners and a surface skim to others. In reality, the course is an introductory basis for continuing study and experience. With the the PDC they can become good permaculturists and more effective designers.

UNIT 2
Principles of ecology

This is a keystone unit. Many of the concepts developed here will be repeated again and again, zones II, IV and V, and in site analysis, especialy grappling with design problems. The concepts work by inserting a wedge to examine a problem; they are a way to start looking at environmental issues.

Humans do not create ecosystems, however they design the conditions under which ecosystems can emerge. Humans cannot disconnect themselves from life. Because they are intimately connected with all life and their actions impact on it; recognition of this is known as the ecological imperative.

In designing and implementing cultivated ecosystems, permaculturists aim to mimic the natural ecosystems in that they:

- Produce no waste through using all resources
- Require only natural maintenance e.g. Birds, animals to fertilise, pollinate and prune
- Efficiently use water and energy and other resource inputs

The first ecologists spoke of closed ecological systems. However reality shows us that we actually have nested ecosystems i.e. each within a larger one. Ecosystems cannot be confined by fences because animals, water and plants do not observe them. We need to design nested ecosystems which means having regard for sector analysis.

Permaculture is based on ecological principles and these will be used later to design abundant, cultivated ecosystems.

The main ecological concepts students need to work with are:

- Food chains and food webs are the basis of ecosystem structure
- The three bio-components of ecosystems are: producers, consumers and decomposers
- Structure is provided through the number of links created by a diversity of species
- Succession can be anticipated and accelerated
- Ecosystem evolution is limited by specific factors
- Stacking occurs in time and space
- Edges, ecotones, intergrade two or more ecosystems and are valuable
- All elements are optimally used in natural ecosystems
- All matter cycles and two that we study are carbon and nitrogen

For two years in the early 1970s, Bill Mollison saw all these working when he retreated into the forest of Tasmania after seeing environmental degradation.

Learning objectives

Permaculture is based on ecological principles. Students will learn basic ecological concepts to use when they design landscapes.

By the end of this unit, students will be able to:

- Describe the flow of energy through a system
- Draw a typical cycle of materials
- Relate these to diversity/pollution/sustainability
- Explain the time/diversity/stability graph
- Discuss the concept of limiting factors
- Explain different types of resources, e.g. renewable, exhaustible

Teaching tools

Use felt board and stickies to enable students to participate in the concepts below:

- Energy flow Through a System
- Cycling of Materials
- Graph of Time/Diversity/Stability
- Drawing of Producers/Consumers/Decomposers
- Drawings of Food Chains and Food Webs
- Drawing of a Mini-cultivated ecological design

Terms

Biomass Total plant and animal matter per unit area by weight.

Climax communities Those plant/animal communities which have reached dynamic structural stability.

Community Any naturally occurring group of different organisms inhabiting a common environment, interacting with each other especially through food relationships and relatively independent of other groups.

Diversity Difference in character or quality, variety, different kinds.

Ecology The science dealing with the relationships of living organisms to their surroundings, habits and modes of life.

Ecosystem The interrelationships of biotic and abiotic factors giving rise to identifiable systems.

Ecotone The transitional area between two or more ecosystems – an intergrade area, also called edge.

Habitat The locality in which a plant or animal naturally grows or lives.

Limiting factors The limit or potential of a system or organism. In this context those which act in ecosystems to prevent them achieving forest status, e.g. wind, rainfall, frost etc.

Monoculture Large-scale plantings of a single crop.

Pollution The action of defilement, impurity, or uncleanness. Excess to any system – not absorbed. Unused resources. Unused resources – surplus to needs.

Stability Resistance to displacement – stable equilibrium – power of resisting change of structure.

Succession Stages of growth from least persistent to more permanent.

Ethics

Maintain and replenish the essential processes and life support systems carried out by nature.

Principles

- Preserve genetic diversity
- Respect the right to life of all species
- Ensure sustainable use of species/habitats

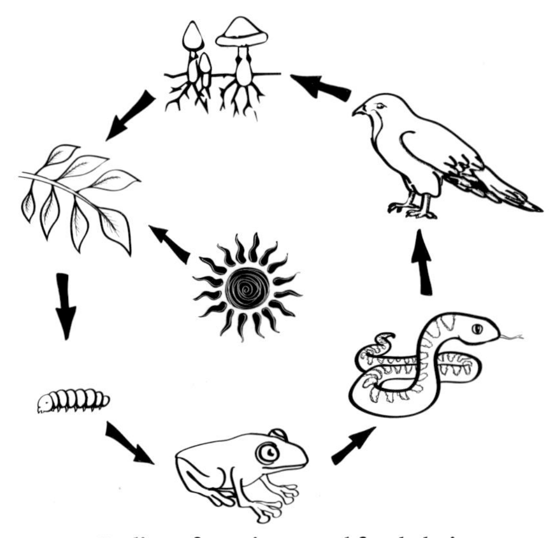

Cycling of nutrients and food chain

Write their answers on the board. Focus on ecology as the basis of permaculture.

Ecology integrates all the above and studies interdependence and inter-relationships. Ecology can embrace the other studies. Ecology normally deals with units called ecosystems.

Ask students if they know any ecological principles.

Functions of ecosystems

Ecosystems carry out functions for life processes. Often the full details of these are unknown or partially understood. Many cannot be replicated by humans. They are invaluable and when we destroy or weaken ecosystems we undermine life processes.

Some ecosystem functions:

- Clean water through oxygenation, wind, plants, soil
- Prune plants with animal agents, wind, insects
- Fertilise soils and so plants through composts, animal manures, nitrogen fixing plants, filter dust, viruses, diseases, toxins through soil, plants aerial systems and ground
- Sterilise with the sun
- Seed dispersal with wind, water, animals, gravity
- Pollinate through wind, animals, insects, water, fermentation

Can you think of others?

Humans are the primary disrupters of ecosystems

Ask students how humans disrupt the following:

- *Soils*
- *Water*
- *Air*
- *Monocultures and vegetation*
- *Toxic substances*
- *Overloading nature*
- *New materials*
- *Reduction of species*
- *Non absorbed chemicals*

Basic principles of ecology

- Flow of energy as carbon, heat
- Cycling of materials, especially nitrogen
- Ecosystem Structure and Function

Biotic components of ecosystems

- Producers
- Consumers
- Decomposers

Ask students to draw a tree and then discuss its function as a producer. Add to this some animals of many types and discuss their functions as consumers. Finally add the soil organisms as decomposers and discuss their role. Challenge students to make the connection between these and the Flow of Energy and Cycles of Matter. Try removing one of these elements and determine what the Earth would look like if any one ceased to function.

Mobile and fixed species in ecosystems

Every ecosystem consists of plants and animals which together carry out all the functions necessary for the perpetuation of the ecosystem. Plants provide organic matter, soil protection, habitat, slow water infiltration and cleansing, food, and shelter while the animals provide pruning, pollination, seed dispersal, fertiliser and cultivation.

Ask students what gardeners do. How can they as designers, replace human work by using plants and animals in their design.

Food webs and food chains

These are the feeding relationships between producers, consumers and decomposers as driven by the Sun and supplied with materials. They constitute the basic structure that enables the ecosystem to function. Any resources which are not fully used can be considered as pollution. The stronger the ecosystem structure the greater the likelihood all outputs will be fully used for productive purposes.

Ask: What is a food chain?

Food chains are inter-connected as two animals may share a resource for a short time. Connected food chains form a food web. Many food webs interweave in a diverse garden. This is the structure of the ecosystem. An ecosystem is strong where there are many links and strong nodes. Nitrogen fixing plants are nodes because of their importance. Some nodes can be destroyed without much damage to the structure but others are critical. They are a measure of resilience. The permaculture design objective is to build strong structures in gardens, food forests and agriculture. Consider a very weak structure of a monocultural wheat field, for example, which has almost no links nor nodes. Even if we don't know all the elements of the ecosystem, by increasing biodiversity we certainly will increase ecosystem strength.

Draw on the board or show posters of:

- *Energy flow from the Sun – encourage discussion from the class*
- *Cycles of materials – show at least two, include nitrogen, and illustrate different timeflows*
- *A diverse and fairly complex food web with sun – solicit class knowledge of structure and function*

Ensure that students arrive at the conclusion that the more diverse and inter-connected the elements, the more efficient, stable, and self-perpetuating the ecosystem. This is one of the goals of permaculture design.

Succession and limiting factors

Draw vertical and horizontal axes to develop a graph.

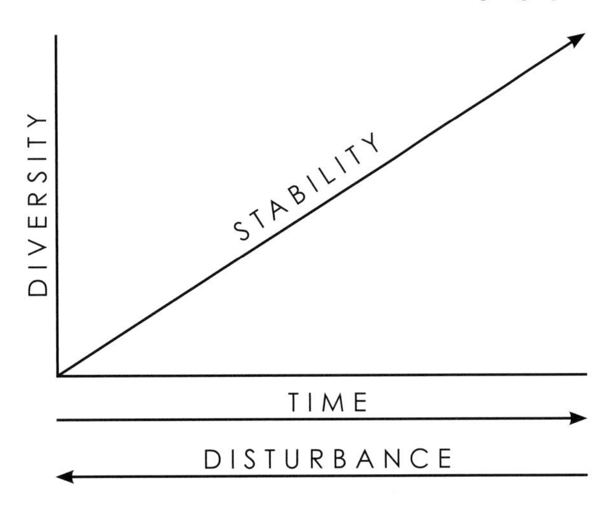

Ask a student to read the following words from the graph axes:

- *Diversity of species on the vertical axis of the graph*
- *Time is on the horizontal axis and*
- *Disturbance reverses time on the horizontal*

Starting from bare ground – if there are no limiting factors every ecosystem will gradually move towards stability which is a relationship between time and species diversity.

The progression (succession) passes through several stages, more or less distinct. These are grass, herb, tuber, shrubs, N-fixers, trees, climbers and fungi and finally achieve a forest. Think of plants as nutrient pumps and stores.

Pioneer plants are often nitrogen fixers preparing land for the final dominant species by changing soil, providing protection etc.

In permaculture, an important design aim is to move to the stability and permanence line as possible, i.e. maximise species and their interactions over time with minimum disturbance. Introduce abundance.

Ask: Why don't all ecosystems end up as forests?

Principle of limiting factors

All living things have needs that have to be met. If all needs are not optimally met, then growth is reduced. One design technique to increase meeting biotic needs and stability is to examine factors that could limit growth, then modify it. For example, consider how water, soil, weeds, pests, sun and cold affect growth rates. Permaculturists modity these to enhance favourable conditions for ecosystem evolution.

In Australia, Eucalyptua species fail when planted out in the middle of paddocks. Discuss what limiting factors might be causing these plants to fail.

Stacking in time and space

The non-use of space costs time and money e.g. bare fields and weeds. With the understanding that each species, plant or animal, provides something for the needs of others, species can be stacked. In a permaculture system, this stacking can be mulch, ground cover, birds, animals, shrubs, bees, canopy species, and climbers. Essentially it results in having yields on multiple levels.

Ask students in groups of two or three to draw a stacked system.

Biotic components of ecosystems: producers, consumers, decomposers

Example of an integrated, sustainable design: Chickens in designed ecosystem

Progressively draw one element at a time asking what the needs are of that element. Build up a picture. Banana tree needs; water, nutrients, weed control, pest control and wind protection. Chicken needs; grubs, water, greens, grains or seeds. Humans can work very hard for the plant to meet these needs. However by designing a cultivated ecology these needs can be met from within the system:

- Banana catches water in leaves and roots
- Clover provides grass/greens ground cover, nitrogen
- Bananas self-mulch and bind soil
- Chicken eat banana beetles, add nitrogen through manure
- Acacias add nitrogen, act as a windbreak and provide seed for chickens

The banana is thus satisfied for all its needs and is also self-perpetuating. The chicken is largely supplied for its needs. Humans do less work.

Student activities

- *Observe and draw a food chain or part of one in your garden.*
- *Describe some decomposers in your compost heap.*
- *Collect a bucket of kitchen organic waste, bury it about 20cm deep. In two weeks dig it up and see what decomposers are there: how does it smell? What has broken down first?*
- *Sit and watch your garden for an hour each day.*

Fukuoka said: 'If we throw nature out of the window, she comes back in the door with a pitchfork.'

THE ENVIRONMENT GAME

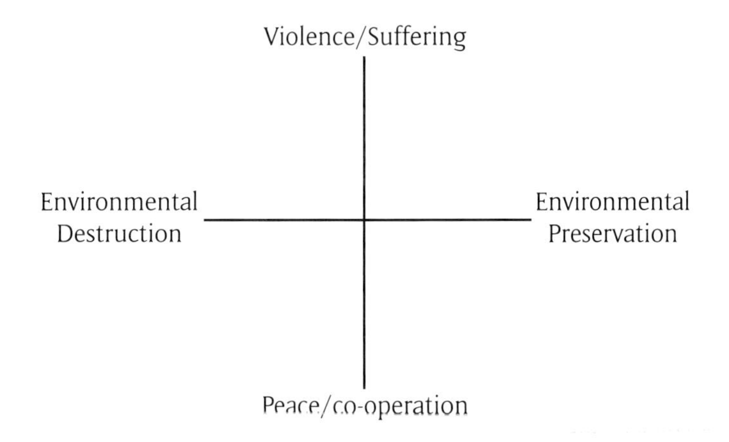

TO PLAY THIS GAME:

1. *You will need as many counters, buttons, stones etc. as there are players.*
2. *Select the first item on the list and the first player thinks about it and then places their button at a point anywhere on the page which they believes is appropriate to the topic. e.g. war goes to the top left hand corner and organic gardening goes to the bottom right hand corner.*
3. *°Everyone discusses their choice and then the next player chooses a topic, thinks, places their counter and so continues the game.*

TOPICS

- War
- Yoga
- Coca cola
- Bicycling
- Private cars
- Cash crops
- Disposable nappy/diaper
- Electric hot water system
- Television
- Commercial soaps
- Plastic bags
- Public transport
- Street theatre
- Peace rally Banks
- Number of Species (Diversity)
- Time
- Disturbance
- ...Add your own items

UNIT 3
Ethics, principles, characteristics

Earth is in crisis. Water, soils and air pollution are just some of the environmental problems we are experiencing, yet economic growth is still promoted as necessary. Our economic and scientific disciplines have been reductionist and without ethics. By contrast, permaculturists consider public and personal ethics to be fundamental to the well-being of Earth and permaculture is distinguished from many other disciplines by its clear set of ethics.

While ethics are seen as a general moral code of conduct, it is the principles which provide a more specific set of guidelines upon which earth repair strategies and techniques are developed. Ethics are moral beliefs while principles apply universally to all cultures and times. Strategies, how and when, apply in time and space while techniques, how, are local. So although few bioregions use the same techniques, they will share ethics and principles.

Select a principle e.g. practice co-operation not competition, and have learners suggest examples i.e. discuss, help each other in learning, or in task work.

Tell your learners the history of the origin of permaculture. It is a good story. After observing a forest for two years, Bill Mollison returned to the University of Tasmania, thinking about what he had seen and, he, with David Holmgren, developed the permaculture ethics and principles.

The permaculture ethics have remained constant since first elaborated although a fourth ethic is sometimes added.

However the principles are more fluid. Mollison gave principles in his *Manual and Introduction to Permaculture*, and Holmgren in *Permaculture Pathways to Sustainability* has devised a series of 12 principles. Morrow listed strategic and priority principles for design.

There are some common principles. The difficulty is that it is not clear whether some principles refer specifically to:

- The landscape or societal analysis or
- The design process or
- The implementation of design

As a teacher you need to give your learners guidance about the principles. I tell the history of the principles and mention that permaculture is still a prototype, only 30+ years old. The fluidity of the number of principles is a measure of the vigour and potential of permaculture.

Learning objectives

By the end of this unit, learners will be able to:

- State the permaculture ethics
- Describe some of the principles of permaculture
- Explain the origins and diversity of permaculture principles
- Discuss the relevance of the permaculture characteristics
- Describe one cultivated sustainable landscape

Teaching tools

- Chalk or whiteboard and pens
- Sketches from *Earth Users Guide to Permaculture* (second edition), your own or others
- Cards with the main principles written on them
- Photos illustrating the principles

Ethics

Develop continuing class consciousness of their need for personal and public ethics and continue to build co-operation and trust.

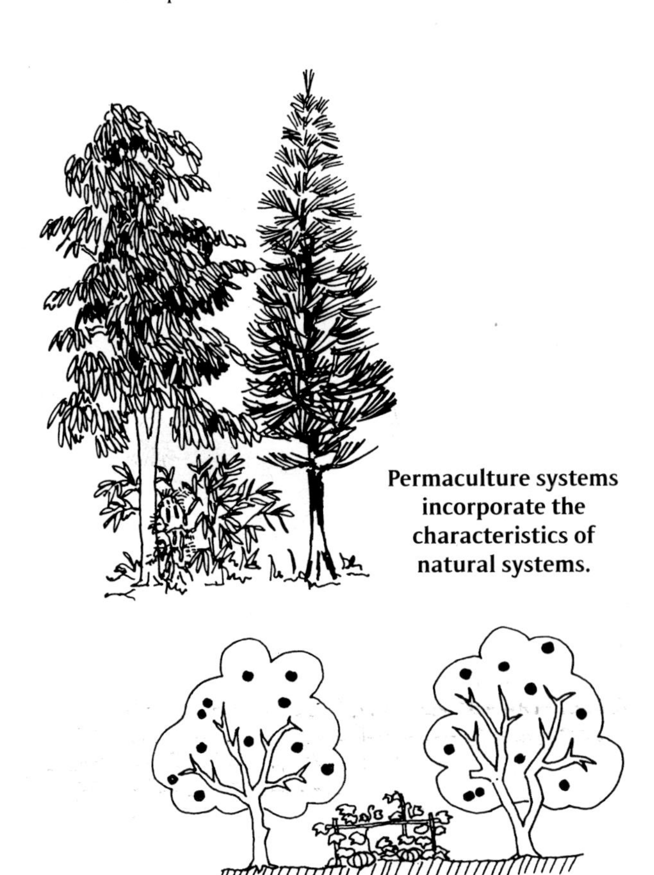

Permaculture systems incorporate the characteristics of natural systems.

Principles

- Work from principles to strategies and techniques
- Use principles to solve problems
- Act locally to impact globally

Terms

Conservation Preservation from destructive influences, decay or waste; use of a resource at a lesser rate than its rate of replacement.

Ethics The moral system of a person or school of thought; the science of human duty in its widest extent.

Intensive Applied to methods of cultivation, fishery etc. which increase the productivity of a given area – opposite to extensive.

Principle Origin, source, foundations, elements, source of action. A fundamental source, or primary element, force or law which produces or determines particular results.

Restoration The action of restoring to former state or position. Rehabilitation – the repair of degraded landscapes, habitats or environments.

Resources Stocks or reserves, organic or inorganic, which can be used by humans for material benefit.

Open class discussion by asking thought-provoking questions which relate to the ethics and principles. Class discussion will strongly reinforce learning this material.

- *Do soil scientists benefit the Earth?*
- *How do ethics affect what happens to earth's resources?*
- *What would happen if we 'distributed our surpluses equitably?*

Permaculture ethics

- Care of the Earth
- Care of people
- Distribute goods surplus to our needs
- And sometimes added: Set limits to population and consumption

Recently Bill Mollison has said: 'There is only one imperative: Take responsibility for your life and that of your children and take it now.'

Discussion

- *Who needs to practise these ethics – us or them?*
- *How would practicing these ethics change our lives?*
- *How can we implement the ethics?*

Mention: Consumption is more decisive than population. It is estimated that it would take 20 billion of the world's poorest people to equal the consumption of the world's one billion richest. Who is using the resources?

Permaculture principles

Permaculture principles are best known for guiding us to design systems that are low impact and ecological balanced. Examples to date most easily demonstrate Care of the Earth. Care of People has fewer examples. I have therefore focussed on this second permaculture ethic and how the principles can work with it. On page 20, I explore Bill Mollison's principles in relation to people, communities and society. On page 22, I offer a chart of principles and potential ways we can implement them in relation to people. On page 23 I explore how the three ethics of Care of the Earth, Care of People and Fair Shares can be applied to David Holmgren's 'Triple Bottom Line', and then apply his principles to Care of People.

These contributions are only a starting point for discussion and are not meant to be definitive. Decide yourself how best to present the ethics and principles. Give examples of each, or better still ask learners to give examples so you can see they understand them.

Below are some early Mollison principles. They are still useful and work well:

- Everything works in at least two ways
- See solutions not problems
- Make things pay
- Use everything to its highest capacity
- Bring food back to the cities
- Help make people self-reliant
- Minimise maintenance and energy inputs
- Maximise yields

Ask:

- *Each learner to select one principle and say what it means for them*
- *Each learner select a photo or sketch and say which principle it applies to*
- *Learners group principles according to Earth or People ethics*
- *You can think of some more activities*

Permaculture characteristics – how it functions and what it looks like

Permaculture landscapes and buildings are distinctive but never the same since the land determines the design and the principle to Work with Nature applies. Photos and actual gardens are very useful. Always give concrete examples as well.

A permaculture design demonstrates the following:

- Intensive rather than extensive
- Diversity in species, cultivars, yields, microclimates, habitats and functions
- Integration of agriculture, animal husbandry, forestry, foraging and landform engineering
- Adjustable to marginal lands
- Makes use of naturally inherent characteristics of animals and plants and their relationships to the natural characteristics of landscape for maximum utilisation, to create an environmentally safe and self-sustaining agriculture
- Use of wild and domestic species
- Long term sustainability

Questions for discussion

- *Could cities raise all their own food?*
- *Should sheep and goats be raised in cities?*
- *What happens to marginal land?*
- *Explain tractoring windbreaks/mulches.*

Observation

All the principles are underpinned by observation. Permaculture works with tangibles. Permaculture observably works! From observation, students can derive empirical data. Students are asked for the first week of the PDC to observe, observe and observe. An outdoor classroom is the best because so much of the natural environment is able to be used as evidence and from which to draw inferences. Good observation is idea laden. Students who 'see and correlate' well will think and solve problems better.

Student activities

- *Discuss with friends and family, elements of your environmentally friendly lifestyle*
- *List how your home environment meets or fails to meet the ethics and principles of permaculture*
- *How will your life change when you practice permaculture ethics and principles*
- *Keep an observation diary.*

Additional information: Comparison of Principles

Mollison's Principles as Care for People

Contributed by Nick Heinamann, Afristar, South Africa

Consider all living species

Relative Location – We are all elements in a system. Our placement and functions in relation to each other is the key to social design. Hmmm.... Household? Community? Village? Society? Based on a web pattern with its nodes and links we operate as nodes accepting and sending our physical surplus and human energy, or, as links which connect the nodes. If the link is too long or the node too isolated by physical distance or personality, it is easily broken. In a social network too few links or too distant nodes mean a weaker system or with many links and closer nodes, greater social cohesion.

Stacking Functions – We each perform many roles in a social system, professionally, socially, and spiritually. Very often we label ourselves one-dimensionally, 'I am a teacher', or, 'I am a single mother/father'. When we are able to see all the talents, skills and uses we bring to our community or society and find ways to make the most use out of these as possible, the more we will flourish as individuals, and the more our society will thrive. Yes, I am a permaculture educator but I am also a yoga teacher, a gardener, a poet, a cook, and I am also a sister, and an aunt, and a daughter. 'Don't put me in a box'.

Redundancy – Each function within a social group should be met by more than one person and create a diversity of roles. For example, in any community there should be not only one healer but several who heal in different ways.

Efficient Energy Planning (Zones) – How and where we work – Mollison says using energy efficiently is also efficient economic planning. Away with the suburbs or transform them completely? Think about the proximity to places we spend our time and its importance. For example, if the core social space we inhabit is the home, but in order to support our families we work 12 hour days in an office 35 miles away spending three hours commuting, then what is efficient energy planning in terms of our social needs? How directly does work feed back in to home, zone 0?

Use Biological Resources – In ecological systems this means making use of plant and animal energy and functions in place of machines. For society I think this would mean using us. Do we prefer bank tellers who we can greet and joke with and ask questions, or the ATM machines? Do we prefer grocers to self check-out? Family and friends to television? Human interaction is healthy; making eye contact, having conversations, responding to the contagion of a smile, or the healing benefits of laughter. That said, there is unhealthy work; toll-collectors, miners, dry cleaners. What other functions could people in these roles do? Or if a type of work is not safe or healthy for humans to engage in such as roadwork or abattoirs, that begs us to reconsider the use and function as a whole and whether it really is providing benefit to society. How can we achieve the functions of a highway without having a highway? The problem is the solution. Think of all the big uglies of society... and then RE-THINK them.

Energy Cycling (Catch and Store) – Use energy as many times and ways as possible before it leaves the system, with the highest possible use first. In society, we can look at money as energy, a sector that comes in a system from a particular direction and leaves from another. Analyzing that flow and employing strategies for keeping that energy in the system and cycling it as many times as possible – local economies, buy local, CSAs, farmers markets etc. Another form of energy in societies is human energy. Can human 'will' be harnessed? If you have the energy, do it? How can this energy be applied... (hmmm?) Let children use their curiosity, ingenuity and sense of play unrestrictedly and employ it – like the play pump which pumps water. Playgrounds that generate power? Create structures that make chores fun or stress-relieving. Combine exercise with accomplishing a goal – the bicycle washing machine, the work-out gym that generates power for the school next door. What is the ferro-cement tank equivalent in society? The think tank? How to capture, store and efficiently use social capital? (See storytelling, seeds and water in 'Catch and Store'...) Perhaps multi-task – 'work where it counts'.

Small-scale intensive systems – Support local governance, community credit unions, bioregions, local currency, farmers markets, freecycle, city repair. Local, community-scale modes of banking and trade social interactions allow for tangible, legitimate membership, true belonging and ownership. The systems will necessarily fit those they serve, as they are 'locally-appropriate', and small enough to be held accountable to their communities. (What does this imply about the internet?) Add to these local schools, clinics and trade schools.

Accelerate Succession & Evolution – Encourage innovation and creativity in schools, workplaces, and homes. Establish venues for inter-generational interaction such as mentorship or seva (service) between the youth and the elderly, big sister – little sister type activities, social gathering spaces, like community halls, that offer the space for storytelling, family art expos, poetry readings for all ages, talent shows. It is necessary to recognize the stage of succession we are in just as it is important to recognize when a forest, society or economy is in its climax phase. Then we either must maintain it wisely, carefully and regularly to keep it there, or we must allow it to burn and rise anew. Learn from the past, don't reinvent the wheel.

Diversity – As biodiversity is critical to resilient ecosystems, so diverse societies are resilient to adversity. This is true in terms of disease resistance and health, and it is also true of resilience to forms of oppression, injustice and disaster. The more diverse a community, the more difficult it is for one race/gender/political party to dominate. Societies need practices to encourage diversity, with the social equivalent of 'crop rotation' making sure that there are term limits on political office, or making it standard to alternate leadership roles between men and women. Structural diversity operates in community groups that function like guilds with people with different strengths and different sets of needs. They support each other and together are more productive than alone or uniform abilities. Cooperation not competition.

Edge Effect, Value the Marginal – Edges are places of productivity where materials are trapped and filtered. How can we take advantage of edge in society? How can we make marginalization an opportunity? The outskirts of cities, bordertowns and suburbs could all be considered places of social edge. Poverty as a social edge? Homelessness? Beggars and addicts and criminals? How can we value them, turn the 'problem into a solution'? Most human settlements lie on ecological edges such as coasts, where the mountain meets the flats, or the shores of a lake or river where we make use of those edges for trade through seaports and riverports, or for resources for fish, drinking water and irrigation water. Edges also feed the spirit of people who choose to settle on coastlines and on mountaintops. But many port towns have no clean beaches. Many edges have been eroded or damaged as a result of human settlement. If we can acknowledge the value of edge then we can choose to protect it and take advantage of all that the edge offers.

The classic example of a negatively perceived edge are: Bordertowns which are centers of trade, and creativity; dialects from which creole or pidgin emerge; also crime, a black market and currency exchange, rich, cultural exchanges, and a portal between two worlds.

Yet edges can also be defined for the purpose of control, to delineate what will be governed. Edges can also be defined by defining the system itself. How do we shift from control and ownership to an ethic of stewardship?

Edges can be used to solve large social problems because as Holmgren says 'the best solutions to problems can be found in places or cultures where the problem is extreme'.

All these edges...disappearing tribes, languages, trades, skills and arts; the very elderly, survivors of disaster. We can learn from them, and value them.

Catch and Store – Just as seed catches and stores genetic information of plants and transfers it from one generation to the next, we can also catch and store social and cultural energy. For example, valuing and recording indigenous knowledge not only for posterity but to create a culture of use of that knowledge. We must reposition value of indigenous knowledge for effective transfer to future generations.

Storytelling is a way to do this. Stories are the seeds of culture, they hold information and worldviews that are necessary to pass from one generation to the next for preservation of cultural knowledge and heritage, and to understand the context in which each of us lives. If we don't know where we come from, we don't know where we are going to. Stories include local information about the environment, the types of foods appropriate for one's genetic makeup, cultural rites and rituals, and so much more. Stories are the water of culture. So let's dig us some swales and spread the stories.

The next tables are draft ideas trying to balance the ethics of Care of the Earth with Care of People. Most of the examples we use in permaculture courses and discussions are about the Earth. We have tried to arrive at examples for the Mollison and Holmgren principles, for people. Then we have applied the permaculture ethics to the Triple Bottom Line (TBL), also known as Planet, People and Prosperity, or Earth, Society and Economics. In the case of the TBL we have added tentative indicators.

Permaculture is the beginning of an ever improving system based on cultivating social and ecological systems. As the discussion paper points out, permaculture has been strong in its first ethic of Care of the Earth with two-thirds or more of the curriculum dedicated to it. The second ethic, Care of People, is less robust, so here for balance we give examples of behaviour which demonstrate care of people.

Examples for MOLLISON'S PERMACULTURE PRINCIPLES of THE 2ND ETHIC: CARE of PEOPLE
These are linked to the Triple Bottom Line with indicators for achieving transformation.

Principles	**Examples**
The problem is the solution	See conflict as opportunity to transform and use anger creatively Reduce consumption to achieve simple living
Least change for greatest impact	Give honest praise and omit criticism Teach children, elders and marginal groups
The yield of a system is theoretically unlimited	No cap on human potential 'the brain that changes' Multi-skill people Increase good relevant information
Everything is connected	Social networks are powerful
Bring in lonely or isolated people	Effective individuals in a group can have big effects Living, breathing models change others
Relative location	Determine where you fit best. Do what gives you energy Put your energy where it is complemented Network knowledge and tasks
Stacking in time and space	Co-ordinate tasks. Plan priorities before implementing Maximise outputs of activities. Cut out some busy-ness Try radical solutions
Stability is the number of beneficial links in a system	Networking, work with like-minded people, develop a thinking community

All major functions are supported by two or more elements	Social network strengths – work smarter, use institutions, get into systems and work on their ethics, especially trust. Add compassion and acceptance Act as links or nodes in your social network
Relinquish power to nature and effective people	Focus on giving rather than receiving Mentor, share knowledge and skills Diminish egos, political statements Trust until you are deceived
Work with nature	Be role models for children and others Embrace all personalities and see potential in them Acknowledge the positive and possible Allow attrition, germination and regrowth of people

Holmgren's principles and the Triple Bottom Line applied to Permaculture ethics

To achieve a restored Earth we need to measure/estimate whether we are making progress and for this we use indicators. Below, each ethic and its characteristics are the indicators for effectiveness.

Earth/ Environment	**People/ Society**	**Prosperity/Fair share**
Support ecosystems Protect local and global environment Increase biodiversity Repair and restore natural and cultural ecosystems	Support neighbours Value customers Work with valuable government services Uphold social equity, democracy, human rights Engage in improving well being of self and others	Contribute to responsible national economic services Provide ethical resources and services to improve quality of life Practice the gift economy
Indicators Reduced mining and fossil fuels Reduced use of persistent unnatural substances Reduced nature consuming activities More achieved with fewer resources Rehydrated, soils, water ecosystems and forests Abundant biomass created Pollution eliminated	**Indicators** Sustained individual and group satisfaction Engagement in healthy sustainable living Training and development for all Health and safety initiatives Democratic alliances with government Basic needs met Engagement in creative leisure Acceptance of difference and diversity	**Indicators** Established responsible, social and environmental values Buying and selling ethically Invest ethically Supply ethical products/services Makes profit Meet basic material needs such as water, food, housing All citizens meaningfully employed

Examples for HOLMGREN'S 12 PERMACULTURE PRINCIPLES of THE 2ND ETHIC: CARE of PEOPLE

Principle	**Example**	**Comment**
1. Observe and interact	Watch, listen respectfully and interact with the group or person	Identify your group's potential problems and strengths Intervene supportively, appropriately Build social knowledge and experience
2. Apply self-regulation and accept feedback	Feed social observations back onto the group process Practice self-restraint, and accept comments graciously	Assist reflexion and insights and group respect for each other

3. Work from patterns to details	Develop the vision and framework and allow the details to emerge	Listen to the visionaries and the detail from people Work the latter's experience into the former's
4. Obtain a yield	Build trust, co-operation, shared knowledge and language People return to the group with motivation and joy	Get good group outcomes Ensure people feel participation/effort is worth it – that their future will be good
5. Catch and store energy	Celebrate and nurture the group: catch human energy within relationships Document success and failure	Allow differences to emerge to enrich the group Maximise personal freedom and thought Push forward not backwards
6. Integrate rather than segregate	Create bountiful inclusive linkages Use complementary skills sets to unlock the greater potential of the group	Work in groups Adapt to different styles, approaches, visions Practice acceptance Listen to the voices of the least likely
7. Produce no waste	Creatively use all the process outputs	Acknowledge and reflect on conflict Transform conflict so it becomes productive Value and respect all input Waste no human potential
8. Use edges and value the marginal	Be open to transformation Adapt structures and processes to emerging needs	Draw in unlikely situations and people Integrate and listen to all Harness the potential in difference
9. Creatively use and respond to change	Look for possibilities as people change Encourage creativity	Adapt and integrate new social knowledge and skills Be free to move on
10. Use slow and small solutions	Allow time for social processes	Allow people time to learn and mature Teach small groups Teach in small bites
11. Use and value renewable resources and services	Nurture renewal of human potential Create the space and take time to manage the health and wellbeing of yourself and the group	Encourage new people with their ideas and skills Assist people to return to groups or revisit knowledge Support people to offer old crafts and skills
12. Use and value diversity	Embrace social diversity in age, sex, opinions, religion, race and knowledge and skills...and anything else?	Integrate disciplines and cultures Support opposites

This material on the second ethic of Care of People was drafted at IPC9 in Malawi with Rosemary Morrow, Nick Heinnaman and Lindsay Dozoretz.

Since then, there have been large and emotional laden discussions on many websites about the Care of People. There have been suggestions for psycho-analytic approaches and a call for a Zone 00 for the inner person. We hope these ideas will prove useful. Such concepts as respect for people have to be translated into behaviours. So we have tried to do this.

UNIT 4
Design methods

Some students have experience in drawing up plans; others have read Mollison's or Holmgren's books. Draw on their design experience. It can also be useful to introduce Edward de Bono and lateral thinking. Site design works best from a multidisciplinary and biosocial approach. Encourage students to work together in and out of class.

Although we are talking about design methods, it is simply to familiarise students with the concepts and not to actually do a plan. They do only the first step, i.e. Sector analysis. The focus here is on observation and analysis. Not only is there is wind across the site but that it comes from the SE and blows during three months of the year and does damage to flowering fruit trees. Observation needs to be integrated and analytic.

There's no such thing as the perfect site which means that most of the design process is problem solving. Design methods which students learn here consist of:

- Techniques which are how they do something, and
- Strategies are techniques over time, and
- Always working from the permaculture principles

A design is a pattern with parts functioning in relationships with each other. And the main subject of permaculture is design. This unit starts students' design thinking and gives them some approaches.

This unit describes nine methods of design and obviously there could be more. Every design will use several methods. Ultimately good observation and imagination, with careful application of permaculture principles will lead to the finest designs. The design methods are in order of priority. If you don't get through them all it doesn't matter but you must get through the first four. Students must know these and their placement on their plans and how to impose them.

Refer back to this unit at other stages in the course, e.g. Site analysis and design assignments.

Learning objectives

By the end of this unit students will have learned some methods which can be used to assess land and to develop designs. These methods integrate rational, creative and/or instinctive approaches.

Students will be able to:

- Evaluate each of the methods and know when to use them
- Name methods they will try on their own property
- Illustrate zone and sector analysis on a sketch plan of their own site

Teaching tools

- Board work for sector and zone analysis
- Sketch drawings to build up the method of analysis
- A packet of cards, about half a playing card size each with a different element on it. Each student to work on their own basic site map
- Chalk for outdoor work on a large plan on ground or wall

Terms

Cradle to grave cost This is the real cost of any item from the moment it is thought of, through its extraction, manufacture and then cost of its waste. This is cradle-to-grave. So a metal paper clip used many times has a smaller cradle-to-grave cost than a staple which is only used once. For plastic products the cost is smaller the number of times it is used. A toothbrush costs less than a biro. It is worth finding out some of these costs. We try to design products and systems in zoning with very low cradle-to-grave costs, e.g. A lettuce from your garden instead of one from another state.

Deduction Inference by reasoning from the general to the particular.

Embedded or embodied energy Everything is made of water and materials. So a manufactured item desired from an ecosystem has an amount of energy and materials required to make it. It will also take energy, materials and water for it to break down and be re-absorbed into ecosystems. We can speak of embodied water e.g. l.0 litre of bottled water has 200 litres of embodied water. It will have a similar energy cost. All industrial, tinned, plastic, frozen foods have high embodied energy and water. All exports have high embodied energy. For example, when we export rice, we are exporting water.

Food miles What is the real cost of packaging, transporting and service items to get them to a market? For example a lemon grown at my backdoor has zero food miles, while one from the other side of the world can be 12,000 miles. We design with low food miles as an objective.

Sectors Those directions outside the boundaries

of the land from which come energies and forces which affect inside the boundaries e.g. Climate elements of wind, radiation and precipitation in addition to pollution, dust storms, cyclones, noise, weeds, floods, wildlife, etc.

Zone An area in a rough radius around a home where elements are placed according to energy use, quantity yields, frequency of required care, and nutrient recycling.

Ethics

Treat nature as an ally and friend whose ways must be understood and whose counsel is sought and respected. Nature is merciful.

Minimum interference for most restorative impact.

Principles

Design methods are used to achieve specific outcomes such as:

- Increasing biodiversity which enables system stability
- Surplus yields
- Sustainable, resilient systems which buffer climate change
- Using surplus from any one element to be used by other elements
- Systems that meet their own requirements and reduce work e.g. Use chickens to clean up weeds/pest. Needs not fulfilled from within the system result in work or pollution e.g. Weeds not eaten by chickens, or excess manure pollutes water supplies
- Achieving climate stability on site

Revise the principles of permaculture

- Work with Nature rather than against it. For example, we can leave weed and pioneer species to provide microclimates, nutrients, and wind protection
- The problem is the solution. For example, a cold wind can be used for its strength and coolness to provide a refrigerated cupboard
- Make the least change for the greatest possible effect. For example, when choosing a dam site, select the area where you get the most water for the least amount of earth moved
- The yields of a system are theoretically unlimited. For example, when you think you have completely planted a garden look again to find a place for another species. Strive for abundance
- Everything gardens. For example, rabbits burrow, defecate, scratch out roots and cut lawns. They provide shelter for other animals in burrows.

The main permaculture design methods are the next four. Your learners will need solid knowledge of these many times throughout the course. So you can introduce them here and then revise them when faces look blank or students say they have forgotten them. It's important they know them.

Sector analysis

Sector analysis involves mapping, in plan view, energies which originate off-site yet have on-site design impacts e.g. Sunpath, cold and hot winds, water flow from outside the boundaries across the land, views, wildlife corridors, fire, and access. Map the areas according to the North, South, East and West noting the hot aspects, the cyclonic rain, the cold winds, pollution effects etc. Put all this on a map.

Ask students to draw their own boundaries in their books. They are working on every influence outside their boundaries which impact inside their boundaries. As you work, get them to put the information on their own page. Explain this is the first step in Site Analysis and is an integral part of the process.

Draw a map of land on the board, put in boundaries and ask students for the following information e.g. Where do winter winds come from? and summer? and spring? Are they hot or cold? Teacher draws in coloured pens to assist learning. Do the same for rain, and sunshine and pollution, and anything else the class can think of which come from the outside and will affect their land which is to have a design.

Ask why they start with a sector analysis and what happens if they don't do it.

Zone applications

Zone application is a design method which applies a master pattern to a site within its boundaries. It is energy conserving and productive. Initially, it is applied by visualising a series of concentric circles to indicate zones of intensity of use.

I have an acronym for intensity or quality of resources used by each zone as: NEWW i.e. Nutrient, energy, water and work. Zone I uses these most intensively and Zone V least as it matures.

- Make lists of elements or items required on the whole site e.g. People, machines, animals, structures, plants, fuels, waste, etc
- Place these elements in zones according to frequency of visits, water, energy use and intensity of management

Zone 0 The house where most available energy is human and also non-renewable. It is a major polluter, often toxic and vulnerable.
Zone I The primary food garden needs frequent work, as is high yielding, keep it small, close and intensive. It can use some animal energy to eat weeds, pests and cultivate.
Zone II The orchard with poultry, reduces work, raises yields still close to house, uses animals to prune, fertilise, scratch, pest control and eggs etc.
Zone III Cropping area such as rice, corn, tea, plantation crops, large animals such as cattle, goats, sheep, deer, alpaca to eat surplus and fertilise.
Zone IV Agroforestry to meet all forest needs for wood, timber, fuels, mushroom, barks, medicines, dyes, large grazing animals care for it.
Zone V Natural systems are created by nature to surround and protect water, soils, and wildlife.

Develop the nearest area first, get it under control then expand. For the moment we don't want a design, just an understanding of the concept. I draw a big map on a floor in chalk and have students place cards on the elements in it to illustrate the zones. When they put down a card of an element such as windbreak, they must justify it.

Slope, aspect, elevation and orientation are used on-site in Microclimate to decide where to place access, water supply, forest, cropland. For example, consider, in cross section, the orientation of the sun to determine the length of eaves and window sill placement for solar access, or areas of shade and reflection or absorption angles of surfaces. This is repeated later in detail in Zone 0.

Observation and deductions from nature

This is a design strategy which starts on and around the site. It is field observation and is often seen as individualistic and unscientific yet reveals dynamic interactions not deduced by any other means. It can reveal problems such as water, weeds, erosion and history of poor land use.

Select one of the above examples and follow it through with the process indicated below.

There is a process to the method:

- Clarify a theme requiring a design e.g. Water surplus, or weed problem
- Where required use equipment for measuring and collecting information on the problem
- Use an experiential approach of applying full conscious senses to details such as site ambience, special sensations
- Make value-free and non-interpretative notes about measurements, experiences etc. For example, seeping water
- Select interesting observations and list your

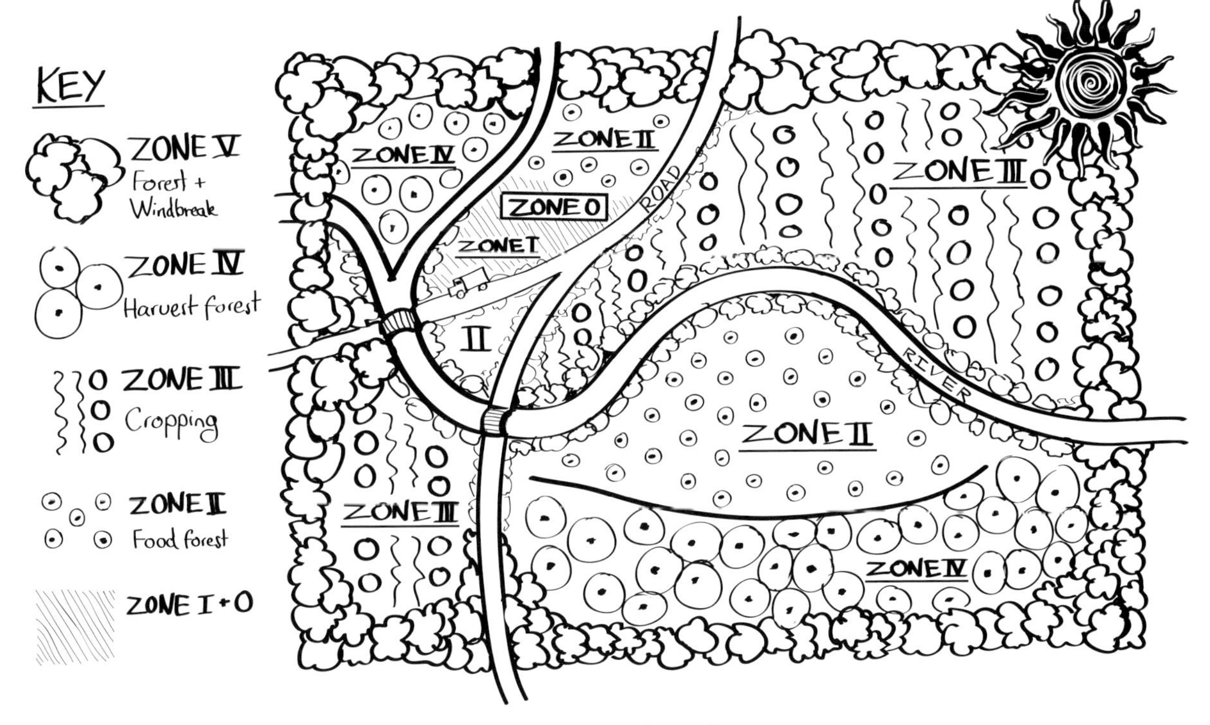

Zones in permaculture design

- speculations such as underground water, house water, run-off areas etc.
- Confirm or deny your speculations by reading, talking, researching, and observing again. e.g. The seepage water could be redirected
- Examine all evidence to hand. e.g. It had occurred two years ago
- Use or incorporate all information by making connections then
- Develop strategies such as directing water to a holding pond and evolve site plans such as water harvesting strategies

Deductions from nature

Good designers seek an understanding of nature. Observations of nature assist in making deductions which lead to good design strategies and techniques. Discuss Masanobu Fukuoka and his observations and deductions.

The deduction method uses all senses with organised, patterned, measured information. Deductions made from nature are used for designing both structures and processes. It is design by analogy; taking the gathered information and using it to imitate or improve on nature in site design.

To arrive at a good design for a windbreak structure:

- Observe the natural placement of woody legumes and natural windbreaks
- Imitate them in a design
- Select a wide range of useful species

To arrive at a good design for a plant propagation process:

- Walnuts in Washington State self-regenerate from seeds falling into valley streams and are carried downhill
- This method is now being used in tropical palm plantations.
- Where else could it be used?

To regenerate large areas of degraded land:

- Birds spread useful and non-useful plants in manure when they roost
- Place posts in bare ground, or build a fence. This will result in a windbreak of desired species and hence a nesting site for predators

---

Ask students for further examples.

Needs and yields analysis

Analysis is the design method used to consider how one item or component can be placed in a design or system, and this component must form a beneficial assembly with the other components in their proper relationships.

The Inputs or Needs, and the Outputs or Yields, of the component are listed to determine where to place the component.

Example: An analysis is carried out to determine the best site for placing chickens in a design. Ask the class for Inputs/Needs and the Products/Yields of chickens. List these on the board:

Chicken Needs (Inputs)	**Chicken Yields (Products)**
Food	Eggs
Warmth	Feathers
Shelter	Gases (farts & burps)
Water	Manure (fertilizer/biogas)
Grit	Crowing/clucking
Calcium	Heat
Dust Bath	Scratching
Friends	Meat/young
Chicken behaviour	Walk, fly, preen, scratch, mate, sit eggs
Breed differences	Dark or light coloured – heavy or light body weight

Draw the chicken house correctly placed so it requires minimum work and resources to meet its needs and all it yields are fully utilised. Look at the chicken house, glasshouse, orchard, pond and vegetable garden to see where to put the chickens in the whole site.

Incremental design – retrofitting

Small changes can be made to existing designs until some optimum limit in efficiency or performance is reached. For Zones of Information and Ethics – read p57 in Bill Mollison's *Permaculture A Designers' Manual.*

Data overlay

This is the classical landscape design technique. A base plan is drawn up and overlays are made of different themes such as vegetation, soil types, water patterns, cultivable land, etc. It is time consuming and has strong pros and cons.

Pros – a good site map makes any landscape design much easier and far more visual. It can indicate sensible options and hypotheses e.g. Potential dam sites.

Cons – map overlays omit minutiae and can never reveal evolutionary processes (e.g Soil creep, regrowth forest).

Some factors cannot be mapped e.g. Ethical and financial culture. The weakness of this method is remoteness.

Options and decisions

This design method requires the designer to think of options for strategies and processes and then to make decisions about including or excluding certain options.

The method can be used for:

- Product or crop options
- Skills and occupations
- Processing opportunities
- Specific market options
- Management skills

It starts with questions or problems. e.g. What can I do in this climate, or what income can I get in 10 years time?

Illustrate one of the above problems using an options and decisions tree. The questions open up options and priorities. These lead to a series of innovative and practical pathways with many potential outcomes. Those that are impractical, unnecessary, unethical or too costly are eliminated. Finish with a few options – it's an uncertain world.

Random assembly

This design process is valuable in assessing energy flows and is also a generator of creativity unblocking problems. This is creative problem solving. It restates the problem in many ways, reverses traditional approaches and allows many solutions to be considered. Very simple solutions can be found by this method.

Process – select all the design components and list them in a circle. In the middle place a random array of prepositions or adverbs. Take a pen and connect them up any-old-how. See what leads to a feasible design solution.

Flow diagrams

This is a design process useful in certain situations, such as efficient kitchen, workplace or time-schedule design. Ask a student to explain flow diagrams.

Student activities

- *Find a design problem and use one or more of these methods to solve it*
- *Report back next class*
- *Observe and deduce seven examples of models in nature which they could use in a design*
- *Try another sector analysis on a different site, perhaps much larger or smaller*

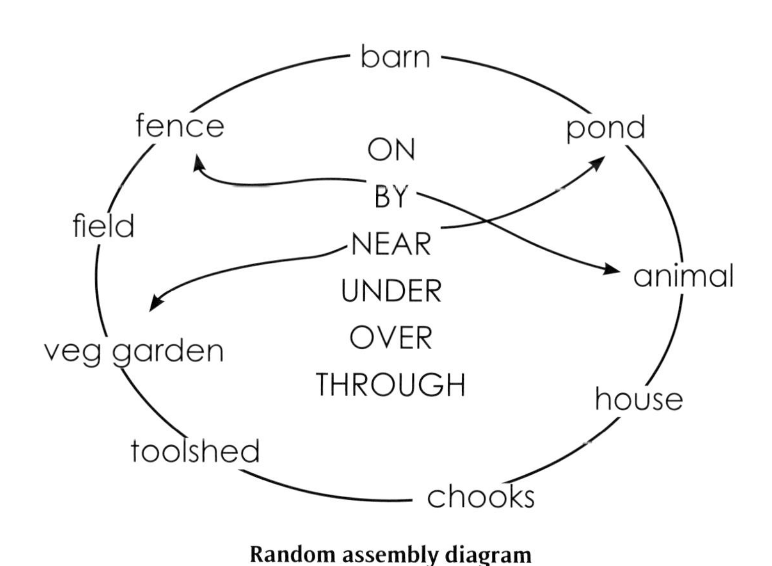

Random assembly diagram

UNIT 5
Map reading

This is a practical unit. Later on, students must look at a map of a client's land and read the topographic features and relate these to the landscape. Maps are one means of gaining information about land potential. There are many types of maps, some are for special uses e.g. Geological and hydrology maps. Maps are part of a whole site analysis approach and working with them can be a design method covered in Unit 4, called Data Overlay.

Maps are very difficult for some students who are verbal or kinaesthetic learners. Yet they are an excellent tool in site analysis. Some students prefer to make scale models, or model land in sand. Some teachers make a paper mâché model which helps those who have trouble with the two dimensions of paper.

Learning objectives

Understanding the topographic features, identifying them, and drawing cross-sections are the useful skills for students learning to design landscapes. In this unit, students acquire practical skills. They learn how to read a map and derive information that is useful in site analysis and design, such as map reading and understanding scale and familiarity with land planning vocabulary.

At the end of this unit students will be able to:

- Define contour lines in their own words
- Show steep and level land using contour lines
- Draw a cross-section
- Identify keypoints and hence sites for houses and dams
- Cross-hatch slope aspects e.g. SW or NE slopes
- Use landscape scales of 1:50 to 1:1,000

Teaching tools

Good board work is useful to build up a picture of a landscape from plan view and from elevation. It is an excellent time to play with sand, plasticine and build landscape concepts. Other aids are:

- Copies of maps illustrating other uses e.g. orthophoto, atlas
- Sand box to make the model in sand
- Model of land taken from a contour map
- Sketches of a cross section, keypoints and site for houses and dam
- Board work showing lines close together are steep and lines far apart are flatter
- Board work showing valleys and ridges
- One A3 photocopy of a map for each student – see at end
- Sheets illustrating keypoint, keyline and degree of slope
- Scale rule to show how it works
- Lands Office or Central Mapping Authority website for state or country. Yeomans Plow Co. for information and leaflets, PO Box 311, Ashmore City, Queensland, 4212, Ph (075) 973 799
- Plasticine, paper and sandbox to make models in either of these

Terms

Cadastral map Maps which show the boundaries of properties and registers property and serves as a basis of taxation (rates).

Contour Interval The difference in contour lines.

Ethics

Recognise that there are unknowns and unknowables when reading sites and maps. e.g. Maps won't show soil pollution or tell of historical uses.

Principles

- Some factors such as historical, ethical, financial and cultural items are often not able to be mapped
- A good map indicates sensible options and hypotheses which can be later verified on the ground
- The scale can be misleading and cause natural features to be exaggerated or minimised. Selection of correct scale is important.

- *Divide class into small groups of two to four students and ask them to discuss the terms listed and where they could find appropriate maps and plans.*
- *Give them 5-10 minutes and list the results on the board.*
- *Show various types of maps eg topographic, orthophoto, deposited plan and show why they are useful.*
- *Explain contour lines, ridges, saddles etc. Ask relevant questions to ensure these are understood.*
- *Work through exercises to gain experience in mapping the keyline system. See p38 and 42 of*

Water for Every Farm.

- *Hand out exercise map and ask students to work with each other to indicate the following features on the map:*
- *North is to the top, where are South, East and West?*
- *Scale and legend*
- *Highest and lowest points*
- *Main and primary ridges, valleys and saddles*
- *Identify key points*
- *Identify and lightly hatch slope aspects e.g. NE and SW*
- *Discuss design implications for desirable house and dam sites etc.*
- *Would you want to buy this land and why/why not?*
- *Demonstrate a cross-section*
- *Ask students to draw a cross-section.*

Student activities

- *Students obtain a topographic map of where they live and show the catchment basin and mark in the keypoints.*
- *Identify where they would put a house and dams.*
- *Measure the horizontal distance between the highest and lowest points using a scale rule.*
- *If they have a larger piece of land, locate dam sites, home site, cultivable areas and those which should be under permanent tree cover – creeks, ridges, and boundaries.*
- *Hatch in a north-east slope.*
- *Draw a cross-section from north to south across a possible house site.*
- *Show students how to make an A-Frame. If one student knows, let them demonstrate.*

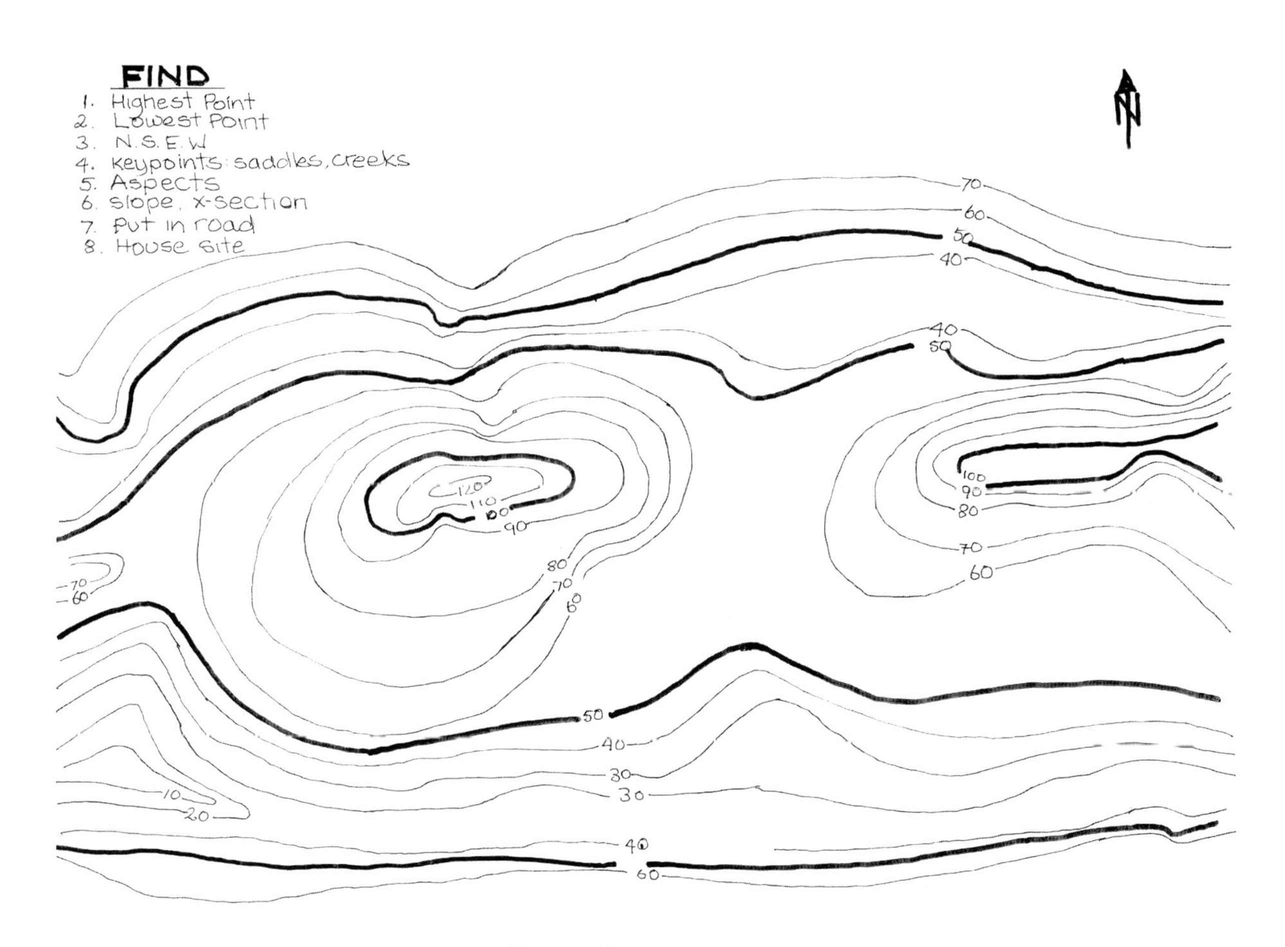

Map reading: contour map

SECTION 3

Ecological themes

UNIT 6
Water and landscape

This is a basic topic and will be referred to again many times. Students need a simple idea of the hydrological cycle of water. They will have some insights from the units on Climate and Soils. These can be briefly revised.

It is important that students understand that:
- The atmospheric circulation of water is the main carrier of heat from the tropics to the poles
- Water is the basis of most biogeochemical cycles
- Water is a primary selection factor in choosing land

Learning objectives

At the end of this session, students will be able to:
- Explain the water cycle
- State the duties of water
- Describe the aim of a designer in using water
- Use several strategies to recycle and cleanse water
- Rehabilitate land suffering from water erosion and/or salting

Graphics

1. A circle showing how much of the world's water is fresh water.
2. A table showing a water audit, i.e. Water coming on to land, stored on land and leaving land.
3. Diagrams of swales and sloping land and the water infiltrating the land.
4. Drawings of distance apart of swales and the slope of the land.
5. Drawings of water cleaning plants.

Terms

Condensation The process by which a vapour becomes a liquid. This process leads to the formation of clouds.

Evaporation The release of water vapour from a surface – the means by which water lying in oceans, lakes and rivers returns to the atmosphere.

Hydrological Cycle The circulation of water from the oceans to the atmosphere, hence to the land, and so back to the oceans.

Keyline The contour line around and across a valley through the keypoint.

Keypoint Point of slope change in a primary valley.

Mulch Material placed or grown on the soil to achieve specific effects – see this unit.

Precipitation The water falling from the atmosphere as rain, hail, sleet or snow. Most of the water which has fallen on the land then returns to the oceans thus completing the cycle.

Rain Seeding A natural process where water droplets form around a nucleus. It is also carried out artificially.

Soil Conditioning The process of re-establishing topsoils and productive soils by increasing water holding capacity (WHC) and organic matter.

Swales Small or large ditches which run along contours and assist land to absorb water.

Some organisms can live without oxygen, but none can live without water.

The functions of water are:
- To act as a medium for the procreation and sustenance of life
- Productive water systems – aquaculture and mariculture
- Energy production – until about 100 years ago water was the principal source of energy eg water mills, canals etc

The total amount of the world's water is constant. Of all the water, only 3% is freshwater and of this proportion, 99.5% is tied up and unavailable. Water cycles rapidly around the earth and this cycle is driven by the sun. It alternates between precipitation and evaporation. They are uneven. More water precipitates than evaporates.

Ethics

Accept responsibility for using water sparingly and maintaining water purity.

Principles

- Use water as many times as possible before it passes out of the system
- Tackle water problems as close as possible to where they originate
- Slow down water flow
- Try to hold all water on the land
- All water leaving your system should be clean

Water is a primary selection factor for land. On a site inspection students will carry out a water audit by listing the following:

Sources of Water on Site

1. Rainfall
- Discuss averages, distribution, intensity
- 80cm per annum and well-distributed is minimal
- Assess the land's ability to hold water
2. Run-off
- Aquire flood data
- Calculate roof capture from all buildings
- Check slopes and soils
- Volume and seasonality of creeks and streams
3. Precipitation
- Dew, fog, snow, condensation
- Rainfall is always wrong – too much, too little, too soon, too late

Discuss silver iodide seeding and tribal rainmakers and forests.

Retain water on land – and keep rivers flowing

Ask students for principles as below:
- Keep trees on all hillsides and slopes over 15-18°
- Keep all river, creeks, springs, forested
- Deep rip and agroplough
- Mulch, keep bare soil covered

Increase water storage – in soil, on land, in biomass.

Soil Storage The most important way to store water is in the soil.

Most soils cannot store water because they are compacted or trees have been removed. Increasing water absorption is the first step in soil rehabilitation. Soil is ripped without being turned over. It is done with any one of three types of plough which are similar in function: Wallace Plough or Agroplow or Yeoman's Slipper Imp Shake-aerator.

Start at the keyline and work uphill parallel to the keyline, not the contour. Over 3-5 years and each year ripping deeper, soil water is increased by 75% and runoff is decreased by 85%. This actually reverses the process of water running downhill. It must be accompanied by tree planting.

Biological Storage Store water in living things e.g. Fruit, milk, eggs, timber, litter etc.

Reduce Runoff Swales – gentle ditches running along contours – slow down runoff and hold water so it has time to infiltrate the soil. If swales overflow then there are not enough of them. Swales use the land as a sponge and distribute water more evenly, holding it on former 'runoff' areas. Swales recharge a depleted water table. On some soils and slopes, after a few years, springs will occur down hill.

The steeper the slope the closer together the swales and narrower and deeper they must be. Plant trees and other vegetation to turn water into biomass, and to retain soil. Swales are site specific and are carefully designed for rainfall, soil and slope.

Decrease Evaporation Mulches are organic and inorganic substances which are placed on the soil for a number of reasons. Ask the class what these reasons might be. Ensure they know the difference between mulch, compost and humus.

Mulches:
- Reduce evaporation
- Inhibit weeds
- Reduce erosion
- Reduce soil temperature fluctuations
- Help water absorption
- Can provide nutrients

Set up the following table – students supply the information.

Inorganic	**Organic**	
Arid Areas	Zone I: Sheet Mulch	Zone II+/ Spot mulch
Tin	Hay (seeds/nutrient)	Groundcover, clovers
Stones	Straw (reflective)	Trees, citrus canopy
Plastic	Sawdust	Trees, casuarina
Sand	Leaves	Litter drop
	Paper (discuss dyes)	
	Cardboard (Boron)	
	Tea and coffee leaves	
	Pine-needles (acid/warm)	
	Compost (nutrient)	

Discuss how each mulch works e.g. Cardboard and paper are weed suppressing.

Rehabilitating Land – essentially water control and planting.

Keyline System of water control places dams on:
Revise Map Reading
- Saddles
- Skyline
- Contours
- Ridges

Each dam has two or three channels in or out. Diver-sion channels can be used via spillways, and lead to irrigation areas. Select keypoints and keylines with multiple dams and channels and irrigate from them.

High dams for: Erosion control, Gravity-fed irrigation, House and stock use.

Low dams for: Irrigation, Aquaculture, Water cleansing, Flood control.

All land work must be accompanied by tree planting.

Soil salinity Swales and dense plantings must be on recharge areas. Use living mulches, deep-rooted trees and perennial crops.

Gully Erosion Stop water at highest keyline before it starts to flow. Slow it down – let it infiltrate. Divert from gully along contours. Plant densely or make a dam.

Re-using Grey Water

Grey water is the used water from washing bodies, clothes, cooking, etc. It can be cleaned by using biological methods so that it returns to natural streams and rivers very clean. Soils and plants are the two keys to cleaning grey water. Discharge to non-food forest e.g. Nuts, oil, wood. Use in glass-houses for methane production and warmth. Pipe through filters to mulched orchards and forests.

- Diagram of biological system
- Diagram of small town or community system.

Grey Water Cleansing

- Aeration by flowforms, windmills, watermills, sun
- Pebble, gravel, sand, calcium, terracotta filters
- Biological – with plants

Plants as Water Cleansers

Plants fix excess nutrients and act as nitrogen and phosphorus traps, e.g. Watercress, rushes, water hyacinth, Scirpus validus (can be weed). Plants desalinate and remove radioactivity, build protein, are low in cellulose, and have 68% digestible protein e.g. Algae. Use of these systems in sewerage lagoons assists in aeration, weed control, attracting waterfowl, and allows cleansed water to be used for urban or community forests especially Zone IV – Structural Forests. For arid lands feed all grey water through underground gravity systems lined with plastic and with plant vegetation on top. Place clean water above housing, discharge grey water below.

Cleansing plants

Juncus spp. (Bullrushes) – take up heavy metals, biocides. They require a seven hour flow-through time, will take up cyanide, thiocyanates, and phenols.

Scirpus spp. – take up heavy metals and biocides, excellent for pentochlorophenyls and other extremely toxic waste.

Schoenoplectus spp. – take up metals, copper, manganese, nickel, kills harmful moulds.

Phragmites spp. – eliminate pathogens, flocculate colloids, dries out sludges.

Econfelder and Associates are world leaders in sewerage treatment – emulsify, diversify and divert sewerage waters to forests and for farming.

Suburban and Urban Sites

People should be selfish about water falling on their land. They need many small water catchment areas, some of these will be for temporary storage. These collect organic matter and high nutrient fine clays, and make good planting places in the future. Instead of draining wet places, use them to grow appropriate plants. A backyard block can have hand built swales which are then mulched over. Try for zero run-off, until soil is repleted, then store surplus on site.

Student activities

- *Ask students to do a water audit of their own homes. FAO recommends 451/person/day*
- *Sydney people now using 455 litres/person/day. For cooking and drinking, people need six litres. Where is their waste? How can they cut down? (Unit 19 – Zone 0 will cover some appropriate technologies).*

Appendix

Many students ask about water filters. Here is a selection.

1. Reverse Osmosis – Unit containing a membrane which lets only fine water molecules, trace minerals and some gases. Expensive, very slow and suitable for people with allergies.
2. Carbon filters – Initially inexpensive but buying replacement filters can become expensive. Buy one with non-filter replacements and bacteria control. Carbon is efficient in absorbing organics. Silver attached to carbon granules inhibits bacterial growth.
3. Ion exchange – Water passes through a column of special resin which removes the salts of heavy metals, fluoride, nitrates and most minerals.
4. Ceramic Filters – Contain a column of porous ceramic material. They need frequent scrubbing.

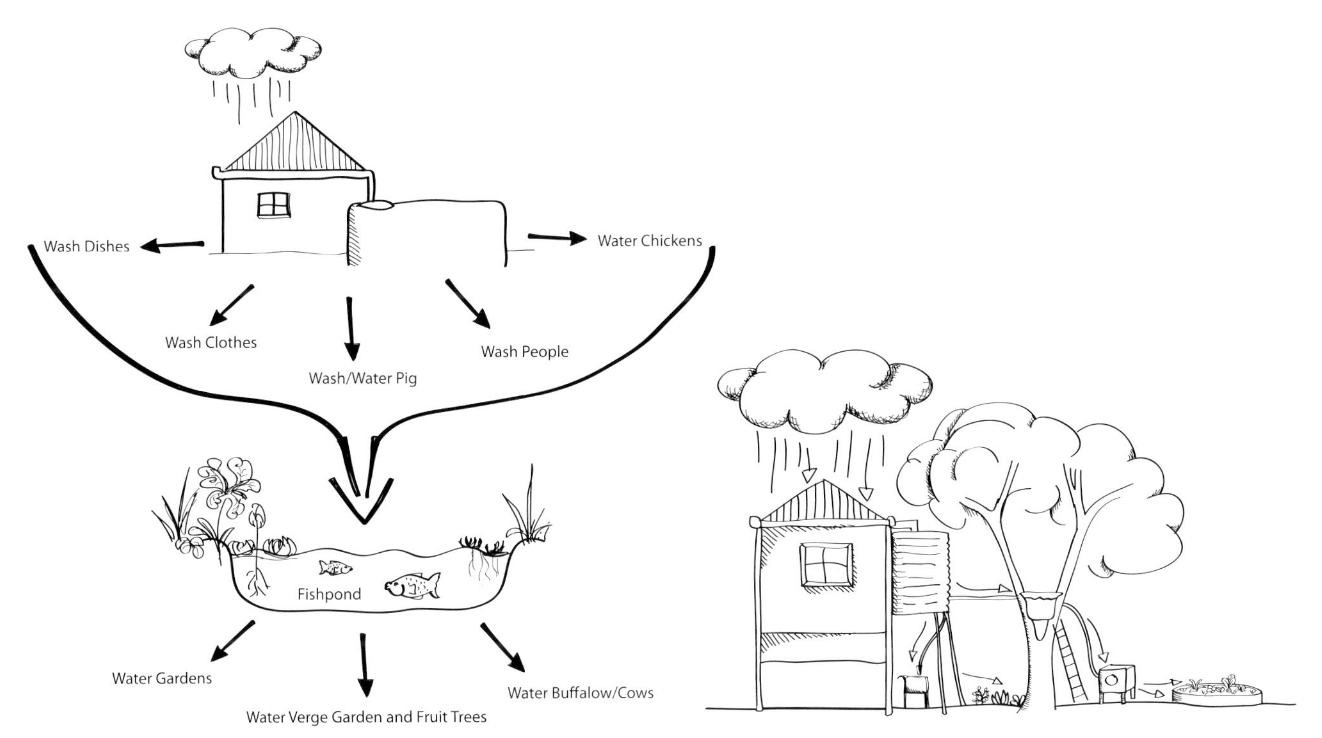

Wet tropics rainwater and fishpond use

Dry tropics water and re-use

Price is not always an accurate indicator of value. Choose easy installation, control of bacteria, cost of replacement filters, if possible a completely sealed unit, Water Board approved, length of warranty and taste and smell of filtered water.

Water in Bali: A system of ritualised ecological management – Water Ecology

For centuries in Bali, water was managed from the top of the mountains down to the coast. The water was monitored by a traditional council who made sure that every farmer and every crop in the guild (or land serviced by the SUBAK, governing body) received the right amount of water at the right time, to ensure a good harvest.

Behind the system was the belief in 'sacred water' and that the gods who dwelt in the water were the purifying agents of it. There were small temples at every point where water was distributed.

The distribution of water was based on a biological view of time, not an industrial view and this coincided with rainfall, with the height of water in the Crater lakes at the top and the needs of each crop.

The symbolism of Bali is in the combination of Mountain + Lowlands, and in Fertility + Water Flow. The full and exciting story of this western discovery of the unrecognised guardian and government of Bali's prosperity is described in the book by Dr. Andre Singer, The Goddess and The Computer. The Balinese had, of course, always known about it.

The Green Revolution took all this apart when scientists decided to produce three rice crops a year instead of the traditional one rice, one corn and one bean crop. The former complicated irrigation schedule for 500 acres of rice terraces could not cope with the increased and almost simultaneous demand for water. The old system collapsed and the new collapsed under an explosion of pests, diseases and water shortages. Now the old system is being partially restored in an effort to lift food production again and reduce the chemical burden.

Extra

1. *List the main water problems in your country.*
2. *List as many solutions that you can think of for these problems.*
3. *What are the three main classes of water and how do you use each.*
4. *How can you clean or keep clean:*
   a. Underground water
   b. Rivers and surface water
   c. Grey water from houses and factories etc.
5. *How many ways can you think of to collect and store water.*
6. *Questions that I have about water are ...*

UNIT 7
Rejuvenating soil

After climate, soil is the main limiting factor to plant growth. It is well established that healthy soils help plants withstand environmental conditions such as drought and pest attack.

All good gardeners and farmers are conscious of their soils. They bend down to touch and smell them. They look at what is growing in them and they exclaim about coldness, warmth, microbe levels, and worms. Eventually, good growers become mildly obsessive about their soils and see them as the basis of all that happens in their garden i.e. pests, mineralisation, aeration, health and fertility.

Most of the soils that students will see, now and for most of their lives, will be damaged, both in cities and in rural areas. Teachers will have to decide what to leave out of this unit since the content cannot be taught in one session. If you feel it is very important, you may decide to take two sessions.

Learning objectives

Students will start to develop positive feelings and attitudes towards soils during the course. So, the objectives for students are to:

- See different types of soils
- Recognise a balance of soil components
- Identify and be able to correct soil problems
- Appreciate soil complexities and plant needs
- Recognise colour, plants, texture, structure as guides to soil, plant and human health
- Develop a real connection with the life of the soil through class theory and practical touching, smelling etc

Terms

Leaching The process by which soluble substances such as organic and minerals salts are washed out of the upper layer of soil into a lower layer by percolating water.

Microflora/fauna Minute to microscopic animals and plants which inhabit a healthy soil.

OM Organic matter.

Peds Aggregation of soil particles to form the structural unit in soil types.

Percolation The descent of water through soil pores and rock crevices.

Permeable Soils which being porous, allow water to soak into them.

pH The power of hydrogen which is a measure of acidity and alkalinity. The scale ranges from 0 (highly acidic) to 14 (highly alkaline) with 7 being neutral. pH in soils measure solubility or absence of an element. pH testing doesn't make it clear.

Porosity Usually refers to soil structure in which peds are full of pores and have minute interstices through which air, light and water may pass.

Salination/salting A state achieved whereby salt is deposited on the soil surface, rendering it infertile. There is dryland and irrigation salting of soils.

Structure How soil particles aggregate together, from amorphous sand through to cement-like clays.

Texture Basically how the soil feels, from silky to gritty.

Tilth The prepared surface soil; the crumb or depth of soil dug or cultivated.

Graphics

1. An appropriate series of drawings to show:
   - Wind and water soil erosion
   - Soil salinity
   - Chemical contamination
2. Table of Traditional Soil Classification
3. Drawings, slides or posters of:
   - Gases
   - Liquids
   - Organic Matter
   - Micro-organisms
   - Particles in soil

Ethics

It is not the purpose of people on Earth to reduce all soils to perfectly balanced, well-drained, irrigated and mulched market gardens. This is achievable and necessary only on the 4% of Earth needed for food production.

Principles of soils

- Soils defy precise treatment as they vary from hour to hour in gases, moisture, micro-organisms, temperature and nutrients, with soil depth.
- The only places where residual soils are conserved or increased are:
- Uncut forest
- Under quiet waters of lakes and ponds
- Prairies and meadows of permanent plants
- Mulch and non-tillage gardens and farms
- Human health is affected especially by the accumulation of biocides and high levels of artificial fertilisers in soils

- The one-off yield of ploughed and fertilised monoculture can outyield almost every other system. But all inputs and losses are not costed. Few, if any, modern agricultural societies do not carry the seed of their own destruction
- The largest job is the restoration of soils and forests for the sake of a healthy Earth
- Without poorly drained, naturally deficient, leached, acidic or alkaline soils, many plant species would disappear
- At a conservative estimate, 60% of the U.K. could be returned to nature

Basic information

What is a good soil?

- A mixture of mineral/particle fractions
- As much compost as possible
- Drains water through in 24 hours, or stays wet 2.5cm below the surface
- Good pores and channels for air to move through easily
- A 'crumb' structure. Explain how it gives away if trodden on
- Earthworms and lots of soil activity.

Now show a graph of a good soil with balance of water, organic matter, warmth. Discuss limiting factors, and how to balance/modify these.

Ask students to bring samples of soil from home. These should be in a clear glass jar with 1/3 soil and 2/3 water and well-shaken, then allowed to settle. Ask students what is growing on the land the soil came from.

Land degradation

The causes of soil degradation need to be recognised in order to repair them. Soil degradation is reflected in:

- Decline in quality and quantity of produce
- Deterioration of natural landscapes, waterways and vegetation.

Write the following eight headings across the board, form small discussion groups to work under each heading. Use powerpoint slides and clear diagrams, especially for salinity.

Ensure most of the following points are covered.

The main forms of land degradation in Australia are:

Water Erosion

Water erosion is the most visible erosion. It is loss of soil through contact with water. It reduces productivity because nutrients, seed and seedlings are lost. The soil has reduced ability to hold water. Large amounts of sediment are deposited in waterways, silting them up and reducing water quality. Water erosion moves up to 200 tonnes/ha/year, (University of Queensland, Faculty of Agriculture).

Wind Erosion

A major problem in dry and overgrazed marginal areas. It is aggravated by drought. Beach dunes are also susceptible. Crop land is liable to wind erosion after the crop has been harvested, or if a fine seed bed has been prepared and the structure destroyed. Loss of tree cover is a major factor.

Dryland Salinity

Caused by overclearing of native vegetation. The soil-water formerly removed by evapo transpiration is added to the groundwater, causing the water table to rise. Where groundwater is naturally saline, the process leads naturally to salting. This process contributes to breakdown in soil structure. Salt poisons the land for normal crops and pastures.

Irrigation Salinity

Occurs because of excessive water use by irrigation farmers, leakage of irrigation channels and inadequate land drainage. Salinity problems are acutely affecting irrigation lands in Queensland, South Australia and the Murray-Darling Basin, for example.

Show maps of salinity increase in Victoria 1982-88 Labelled White Death.

Soil Acidity

By using some type of clover and phosphate-rich fertilisers, the acidity of many of Australia's fragile and infertile soils is increased. Crop yields are reduced by about 50%. The reduced plant cover contributes to wind and water erosion.

Soil Structural Decline

Farming techniques are the main cause of decline in soil structure. The heavy and aggressive use of farm machinery breaks down soil particles, causing problems such as compaction. As a result, crops and pasture yields decrease and soils become prone to water erosion and lack of water absorption. Losses of $300m per year result. This reversible problem is presently Australia's most expensive form of land degradation.

Mass Movement

Occurs on steep or sloping land which has been cleared of vegetation. The soils become unstable

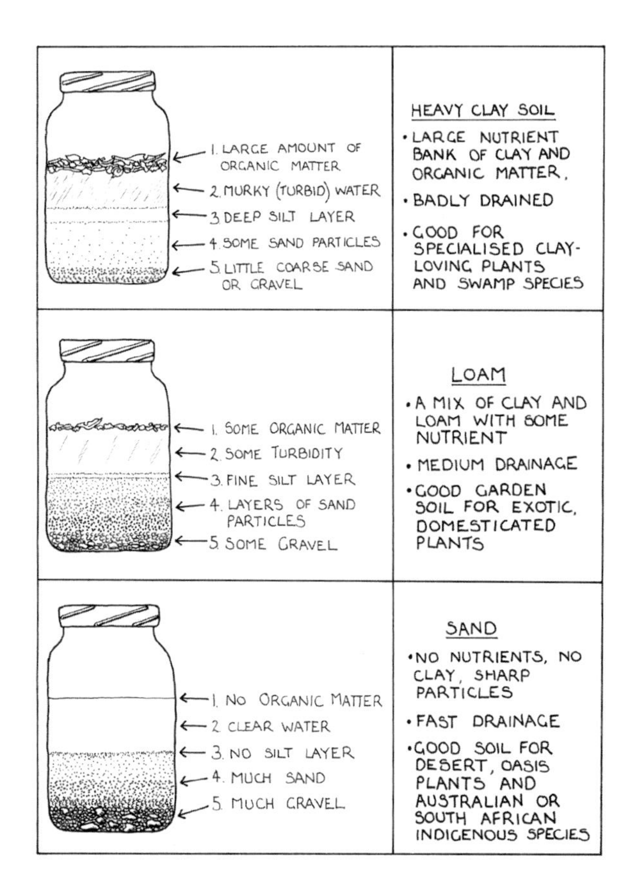

and following rain, the soils move downhill. Damage occurs to houses, roads, railways and farm land. We lose soils at an ever increasing rate. Australia has at most 30% of its soils in fair condition.

Chemical Contamination

Bio-accumulation of substances and biophysical changes take place in soils and water. The chemicals on rural land are mainly fertilisers and pesticides. Some of these are taken up by produce (plant and animal) and are toxic or near toxic. In some cases, the effects are currently unknown. The main effect on soils is the absolute loss of soil micro-organisms operating the cycles of matter and loss of healthy organisms such as natural fungi and antibiotics. Gradually organic matter is lost and soil structure starts to decline.

Polluted soils exist where conventional orchards, sugar cane, pineapple, cotton, tobacco, bananas, potatoes have been grown. These soils should be used only for forestry. They contain lead, arsenic, copper and biocides. For cadmium and radioactivity, the lock-up time can be thousands of years. For areas which have been farmed from 1950 to the present, the costs of rehabilitation may be prohibitive.

Applications

Soil Classification – Tribal or Traditional Methods

These are the methods people have used to tell about soils and don't need a laboratory. They integrate all these points:

Colour	Taste
Slope	Animal indicators
Usage e.g. salt/clay	Elevation
Work	Organic content
Plant indicators	Sand content
Moisture capacity	Holding texture
Structure	Wet-dry behaviour
Vegetation	

Ask students in small groups to discuss 2-3 of each of the above topics. Set up a table on the board after each group has spoken and ask for further information. Add theory where it is required. Characteristics from – to.

Colour

white	*grey*	*yellow*	*red*	*brown/black*
low O.M	no O_2	Al & Fe	Fe, poor	O.M.

Taste

sour	*sweet*
anaerobic/acid	aerobic/alkali

Ask students to carry out a traditional classification on soil from home and bring samples next lesson/ week. Explain and discuss briefly: soil texture and soil structure.

Main Soil Factors

Set up groups to discuss the following topics. Add any points that aren't covered.

Soil Water

Water is the most important factor in soils.

Water is the universal solvent for soil nutrients/ plant uptake:

- It moves down by infiltration
- It moves up by flooding or capillarity
- It moves along slopes by throughflow

Water acts biologically:

- Many microfauna/flora live in it

- Many species concentrate minerals in it
- Is held by humus in soils
- Is dispersed by transpiration
- Carries nutrients from rain
- Is concentrated in metabolic processes
- Concentrates nutrients locally

Water is:

- Bound by clays
- Bound 25 times more by humus in soils
- The solvent for plant nutrients
- Biologically cleaned through healthy soils

Soil Biota

This includes all the organisms in the soil. They form a large mass within healthy soil. Their growth rate depends on the rate of turnover of the mass of food (organic matter) in the soil, and available temperature and water.

Gases in Soils

Sun, moon, tides, winds, air pressure and activity in soils exert changes in air gases. Gases diffuse in and out of soils. Soil processes affect ammonia, carbon dioxide and methane. As plants breathe, so do soils. This is not well-known. Plants trade both ways with gases in soil and in air.

Biological exchange takes place when groups of plants such as algae, rushes, crops, trees and herbs, transpire oxygen, water vapour, hydrogen, chlorine and carbon dioxide. Legumes and algae release ammonia. Root gas exchange is affected by rate of transpiration of leaves. Cultivation methods such as ploughing and earth turning affect a net loss of nutrients in many ways, one of which is atmospheric pollution.

Organic Matter

Organic matter is any material in the soil which has been living or is living, and includes excreta, exudates and slough-off. It is eaten by soil biota and can be leaves, peats, sponges, fruits, timber, bones, skins, fibres and throughfall washed off plants e.g. Insect and animal excreta.

Ask what is the difference between humus and compost?

Mineral Particle Fractions

Soils comprise varying fractions of sand, silt and clay. These have an effect on water holding capacity, nutrient banks, wetting and drying of soils, warmth, acidity/alkalinity, drainage, and work.

Put a line on the board – a continuum. Ask students to list traits of:

- *Clay*
- *Loam*
- *Sand*
- *Water logging*
- *Drains fast*
- *Holds nutrients*
- *Leaches*

Ask students to look at samples. Discuss colour of water, clay content etc. Discuss fractions – organic, clay, silt, fine sand, coarse sand. Revise the following: Texture, structure colloids.

---

pH is a measure of the acidity/alkalinity of a soil. It is largely regulated by organic matter. The scale of pH is from 0 to 14, Acid to Alkaline. The units are multiples of 10. pH simply reveals the available nutrients in soil. Thus its value is that knowing the pH, you know what plant nutrients are available. Alkaline soils are usually nutrient rich, however, zinc may be deficient in these soils. To increase alkalinity add lime (quick results), or add dolomite (slower + magnesium). There are alkaline ecosystems, that can be imitated in home gardens. Acid soils are leached soils, wet soils, and tropical soils. Special plants grow in them.

What pH?

Soils can be made acidic by addition of sulphur. The ideal garden range is from pH 5.5 to 7.5. Different plants have different mineral requirements and so grow in different pH soils. It is good to have different pH areas in gardens.

Show or offer chart of pH preferences of common plants.

Emphasise that observation of colour, plants actively growing, and their disease resistance is as good as, if not better than, relying on pH meters.

Ask what bracken, nettles, banksias, thistles, reeds, blackberry, redgum, yellow gum and even tree habit will tell about soils. Ask for other examples of plant indicators.

Student activities

- *Ask students to tell you what is the pH of their soil deduced from the plants growing in it*
- *Ask students to map soils in their gardens – find at least two types. Students to keep a soil sample*

for one year, try to improve the soil, in the place they took the sample from. Then take a second sample. Try this for three-four years. It takes this long to really improve a soil

- *Find a difficult soil and explain how it can be remedied*
- *Carry out a traditional soil classification*

Additional information

To be taught to an advanced class if there is time, or if you feel it is necessary.

Land degradation

- Refer students to page 185 – Mollison, *Designers' Manual*
- Lead toxicity
- Groundwater
- Persistent Biocides

Soil biota are generally classified as:

- Microflora/fauna – tiny invisible bacteria, fungi, virus etc.
- Mesofauna – to 2mm, mites, termites etc.
- Macrofauna – 2 to 20mm, woodlice, centipedes
- Megafauna – 20+mm, crickets, moles, rabbits

Very small organisms basically lead an aquatic life in soil. They are vulnerable to death or encysting under dry conditions. They may go on strike! Larger animals are confined to pore spaces and burrows.

Soil biota themselves form a reserve of otherwise easily leached nutrients (nitrogen, sulphur etc). They also gather, store and concentrate nutrients. They breakdown organic matter to nutrient forms which can be used by plants. They form humus which stores 25 times the amount of nutrients of clay. Biota form a food chain so that meso- and mega-fauna feed on microflora and finally it is the bacteria in the gut of earthworms that provide the nutrients for plants. A soil rich in micro-organisms (MOs) will eventually provide a balanced soil medium for most plants.

Phosphate deficient soils In Australia, for every 10 tonnes of super-phosphate applied to soils, nine tonnes are insoluble. So insoluble salts build up in soil. In *New Scientist* 9/8/1988 it was reported that a mould, a penicillium bilaji, occurs naturally and makes phosphate soluble so it can be absorbed more easily by crops. This fungus can also stop phosphate fertilisers degenerating into insoluble forms once they have been applied and can also transform natural phosphate in the soil into soluble form. This makes good use of native phosphate. It probably occurs naturally in compost and so applications of compost can inoculate soils. The work was carried out in Canada and has enormous implications for Australian farmers.

The Ethylene Cycle involves organic matter, water, and soil biota.

1. Aerate soil with shaker/aerator plough
2. Microbial activity is stimulated by increased oxygen
3. Micro-organisms increase organic matter and minerals
4. Increased plant growth (sugars and leaf drop)
5. Soil Biota multiply and use up oxygen
6. Their dying produces ethylene
7. A preponderance of ethylene produces nitrate from nitrite (nitrates are very soluble plant major nutrients = protein builders)
8. A large amount of nitrate inhibits ethylene
9. Plant's growth increased by nitrates
10. Increase in soil biota to use up plant and root surplus

Humus is the partly broken down organic matter. It gives soils their rich dark colour and has a major effect on soil properties, structure and water holding capacity. Good soil/humus has a good smell and an almost oily feel. Humus has a negative charge. It absorbs water like a sponge. It acts like a nutrient bank since nutrients adhere to it at 25 times the rate of nutrient storage in clays.

It can be destroyed by burning or is preferentially washed/blown away by erosion. Humus improves the texture and structure of clays, and of sandy soils.

Ask students how it does this?

Humus is added to soils through compost and mulches. It is being systematically removed from soil with modern technological commercial farming. Humus solves pH problems. When composting, the smaller the particle size, the more quickly humus is created. A mulcher can be useful.

Plants contribute to humus as they recycle 25% of their roots each year. The area around the roots is known as the rhizosphere and much of what goes on here is little understood except for the fact that it is rich in activity.

Texture How a soil feels from silty to coarse. It gives information about soil particles and organic matter.

Structure How it holds together. Total clay or total sand is structureless. Think of beach sand, hard clays.

A loam is a balance between clay and sand, air and water move through easily, soil holds together in peds.

Colloids Stable, water gels, or suspension of clay and organic particles in a finely diffused water state in soil. They have a negative charge and so attract and hold positive particles such as nitrates, sulphates etc. Colloids made of humus are 100 times more effective than clay colloids.

Plant Nutrients These are usually classified as macro- and micro-nutrients. Those mainly spoken of are:

- Nitrogen most important for protein synthesis
- Phosphates for cell growth and strength
- Potassium for flowering and fruiting, known as NPK

Phosphates may contain large quantities of cadmium and lead in Australia. Tasmania is considering testing plant products in which heavy metal levels in the soil are close to permissible levels (*Organic Growing*, Summer 1988, p21).

The turn-over of nitrogen by earthworms exceeds that of litterfall of plants. Nitrogen fixing plant species can fix 370lb of available nitrogen per acre per year. It is the actually Rhizobium bacteria in the nodules of roots in leguminous and other plants that fix the atmospheric nitrogen into forms which can be used by plants.

In Australia, termites in dry areas replace earthworms as the recyclers of plant nutrients in the soil. Their nests can be vast underground stores of compost which are available to the surrounding ecosystem when conditions are right e.g. After rain, or in periods of environmental stress.

Other sources of natural fertilisers are animal manures. Every system has its animals as maintainers by dispersing seed, pollinating, manuring, scratching and aerating the soil for the plants. A general rule of thumb is that the strength of animal manures is related to diet. Hence, herbivores such as horses and cattle have a lesser nitrogen content than say, poultry or pigs. Manure from the former is less likely to burn plants if urine is mixed with it.

Difficult soils It is not always possible to select the perfect soil for intensive food production. Some difficult soils are:

- Alkaline – deserts, coasts; use special species such as carobs, mesquites, prosopis, locusts
- Acid – wetlands, bogs, high rainfall, uplands, siliceous soils; use oaks, pines, blueberries, brambles, strawberry
- Heavy clays – lowland, impermeable soils, use gypsum and humus
- Non-wetting soils – drylands (due to fungus-algae producing wax) mulch, mix with clay, or commercial gel

Chemical Additives

- Gypsum – a claybreaker does not alter pH
- Dolomite – changes pH to alkaline, breaks clays
- Bentonite – a wetting agent, expensive
- Lime – changes pH to alkaline, low solubility
- Sulphur – makes alkaline soils more acid

Extra

1. *What is a soil? What is special about a tropical soil?*
2. *What is a good soil?*
3. *What are the three broad categories of soil?*
4. *How can you improve a soil?*
5. *How quickly can you improve a soil?*
6. *What are the five components of a soil?*
7. *Questions I have about soils are ...*

UNIT 8
Designing with climate

In this unit and the following on Microclimate, Soils, and Water, take the same approach. This is to divide up the topic into its elements so they can be studied for the usefulness of each element. Or to see if we can influence it.

In this unit on Climate, firstly we study how climate is madeby the Earth's Rotation and the Earth's axis – and how these have their effect on winds, heat, radiation and precipitation. These last four, Wind, Heat, Radiation and Precipitation are examined to see how to incorporate them into site designs. Elements of climate separately and together affect seed germination, plant blossom, pollination, seed dispersal, fruit ripening, local pests, and other such factors.

Learning objectives

From this unit, students will appreciate that climatic factors are the main determinants of the plant, animal and structural assemblies. Using knowledge about climate, student designers will be able to:

- Implement strategies that spread risk of damage due to climate extremes
- Understand how permaculture designs can modify climate
- Modify and/or design homes that are energy/ water efficient
- Select appropriate plant and animal species
- Effectively use climate elements in design e.g. radiation

Graphics

1. Powerpoint slides or board work of the world with the Hot, Temperate and Cold areas on it
2. Powerpoint slides or board work showing how winds move across the Earth's surface and bring monsoon rain
3. Powerpoint slide or board work showing how the sun's white light heats the earth
4. Diagrams of radiation and effects of heat and light and dark colours.

Terms

Absorption Heat and light taken up – most perfectly by a black body.

Albedo The proportion of solar radiation falling on a non-luminous body. The albedo of the earth is about 0.4 i.e. about 40% of solar radiation is reflected back into space.

Precipitation The deposits of water in either liquid or solid form which reach the earth from the atmosphere.

Radiation Radiant energy is constanly emitted in all direction by the sun, some of this reaches the earth and is converted into heat. The earth is constantly losing heat into space by radiation. Land loses heat by radiation more rapidly than water, and high ground more rapidly than low ground. Heat given off by a radiant body can be stored.

Reflection Light and heat transmitted away from a body, most perfectly by a mirror – next by light coloured objects.

Climate variation is increasing and the variation will lead to increases in floods, droughts, crop failures, temperature extremes, intense winds etc.

Climatologists often do not consider the effects of the removal of forests, of pollution, of cities, agriculture and albedo. Many weather statistics are inadequate due to montane effects, energy transfer from wind and currents, dew, long-wave radiation, UV incidence and gas composition.

Design strategies work by spreading risk by mixing crop species and varieties. For homes, risk and cost are minimised using thermal mass, insulation, good siting, water storage, windbreaks. There is no substitute for local deduction and observation.

Ethics

Strive to maintain climatic stability by designing sites with a maximum of tree species and without risk of polluting the atmosphere.

Principles

There are intimate interactions between site factors such as slope, valley, coastal proximity, altitude and local climate. These interactions are complex and continue being complicated by:

- Orbits of the Earth, moon and sun
- Changes in gases due to vulcanism, pollution, agriculture etc
- Extra-terrestrial factors e.g. Meteors, high level jet streams

Climate factors have their most profound effect on the selection of species and technologies for sites, and are thus the main determinants of the plant, animal and structural assemblies.

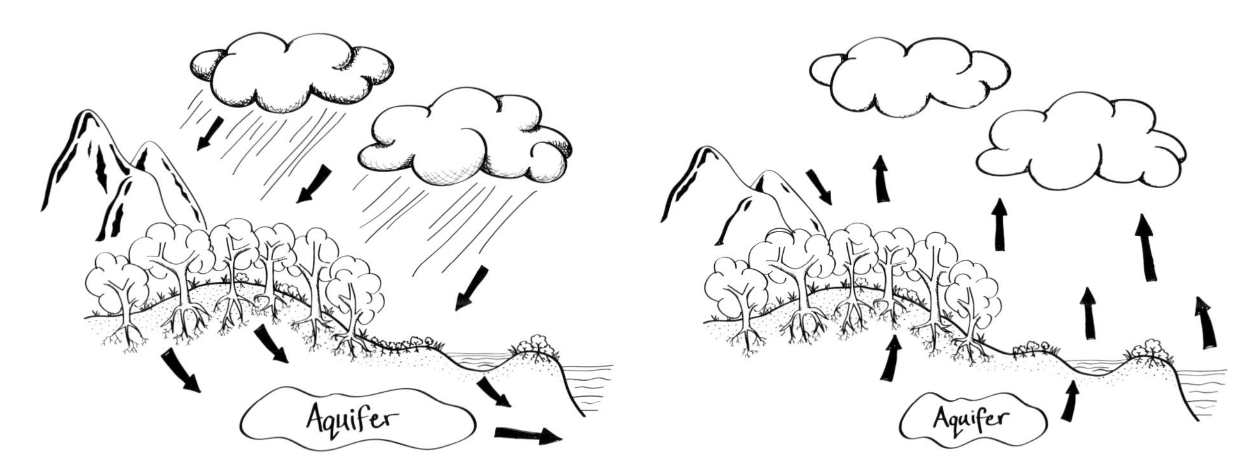

Water cycles constantly between land and sea

Revise the broad climatic zones. Show how these zones are roughly patterned and symmetrical around the globe.

Patterns in Global Weather

1. The Earth rotates around the sun from West to East.
2. The Earth swings on its axis in six-monthly cycles.

These two things have their effect on: Wind, heat and radiation, precipitation

And these three, singly and together, have their effect on: Vegetation, animals, soils

All of the above are modified by altitude and latitude.

Wind Patterns

- Explain winds
- Explain to class bioregional phenomenon such as monsoons in tropical areas

Rain Patterns

- Orographic rain
- Cyclonic or frontal rain
- Convectional rain
- Other Precipitation: Dew, condensation, fog

Radiation Patterns

- Two types of radiation – direct and diffuse. Only direct is reliably measured
- Ozone screens out harmful UV light. Moon and starlight are effective although a weak light source
- Light and heat are elements of radiation

White Light

White light entering the biosphere has no heat as it is a poor quality short wave. When it meets any sort of obstacle e.g. Earth, oceans, buildings, plants, it becomes high quality, long wave radiation and becomes heat energy.

Only this way does it heat the air. It is very difficult and expensive to heat air, as air stores little heat, and so heating air is very inefficient. Oceans are warmed and the warm humid air rises and the warmth is distributed via the hydrological cycle over the Earth's surface (See Unit 6 on Water). It is the moisture in the air which absorbs the heat, not the dry air.

Hot/warm air expands and this can create currents, which can be used in strategies to alter microclimates: e.g. Ceiling fans bring high, warm air down into a cool room.

Heat moves from warm bodies to cooler ones. Never vice versa. e.g. Heating a kettle of water over fire.

Absorption

Light is absorbed, converted and re-emitted as heat. Soils can absorb heat to 50.8cm (20in) depth. Soils lose heat more slowly than air. Soil is coldest just before dawn. This causes frost. It forms on high valleys and ridges.

A black body is a perfect absorber as it converts all absorbed light to heat. This can be radiated back, e.g. Wood heating stoves. The Earth acts mainly as a black body.

Reflection

Incoming light is turned away almost unchanged. A perfect reflector is a mirror with 100% light reflected. White objects are good reflectors. Albedo

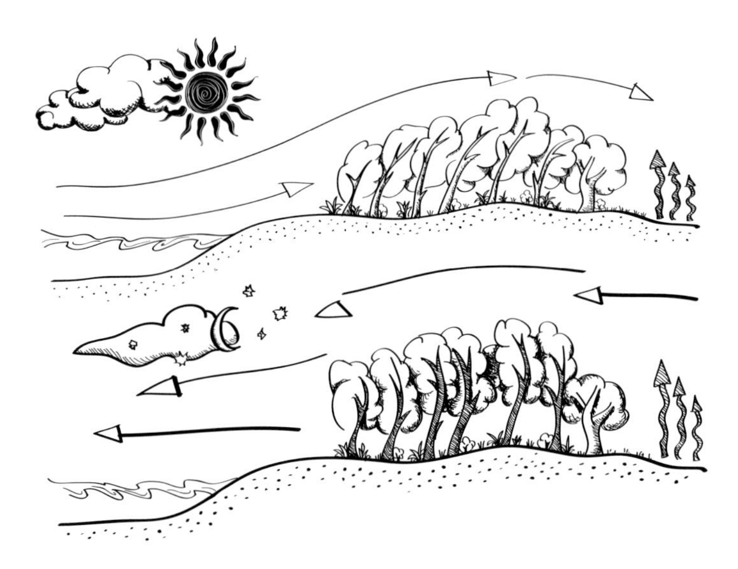

Water masses modify the climate daily (diurnal effect)

is a measure of reflected light value. Reflected light goes back into the atmosphere or is absorbed by nearby surfaces e.g. Off a white wall onto leaves.

Refraction

Light can be bent or curved e.g. Passes through water.

Use powerpoint slides showing fate of solar light (white light) and its losses before reaching earth.

Heat Storage and Transmission

Class discussion of convection, conduction and radiation.
Convection is..............
Conduction is.............
Radiation is.................

Effects of Elements of Climate

Elicit students' experience and ideas about how various elements e.g. Light, cold, photoperiod and temperature can act as limits or facilitators of plant and animal growth.

Light Carrots need a fixed amount of light and thus should be planted shallowly. Some weed seeds and desert seeds germinate only in darkness.

Air Temperature Temperature effects can be either direct, reflected, and/or diffuse. Banksias and Wattles need heat for seed maturity/germination. Some plants act as heaters and store fat and heat to attract pollinators. Temperature can bring forward or delay ripening. It has a large effect, hot or cold, on animal health and appetite.

Cold Stratification or vernalisation means some plants may require a period of cold e.g. Apple fruit set, apple seed, bulbs, wild rice.

Frost is intense back-radiation from soil and some farmers believe it is cleansing by forming soil structure, killing some pests and germinating some seeds. Usually it kills some plants, as the frozen water in plant leaves expand under morning sun. The still air forms frost. Hence the air movement via specially designed windbreaks prevents frost. (Refer to Unit 13).

Soil Temperature An increase in soil temperature can reduce germination time by one tenth to one quarter of that in cold soils – hence use of glass houses and bottom heat. Temperature often limits germination e.g. Celery not greater than 24°C and beans not less than 10°C.

Photoperiodism is the response by a plant to length of light. It affects seed set, blossoming, fruit set, and plants have specific daylength requirements epecially at the change of seasons in temperate climates. Some animal fertility cycles are affected.

Photosynthesis The amount of photosynthesis doubles for every 10°C rise to a maximum of 25°C, when the plant shuts down for the day unless it is shaded. 15°C is the optimum soil temperature for most plants. Low CO_2 can reduce photosynthesis.

Photosynthesis effects plants which absorb light of all different wavelengths. Different parts of plants absorb light differently. Plants in different biozones evolve different light absorption organs e.g. In the tropics are dark green and red leaves, and in the hot deserts there are white barks or light leaves.

Continental Effects Inland extremes and widely fluctuating climatic zones, which are not buffered by sea currents, lead to specialised ecologies.

Respiration There is a 2-4 times increase in respiration for each 10°C rise, and a corresponding rise in water requirement. If stressed by heat, stomates (plant water loss cells) are shut by midday and wilting occurs even though there is sufficient soil moisture.

Climate Modifiers

Ask how climate is modified by the altitude, latitude and aspect

Aspect Easterly aspects are usually cooler and get morning sun etc.

Altitude is not the same as latitude since at a higher altitude the air is rarer, radiation is higher and air pressures are less.

Latitude High latitude areas have long summer days for vigorous summer growth. Severity can be modified by currents e.g. Scotland, Alaska.

Student activities

- *Each student to summarise the climate they live in and the likely effects on biomass*
- *Each student to consider their type of heating or cooling and how efficient it is (major discussion on this in Zone 0)*
- *How does climate affect such things as planting times – discuss lettuce, cabbage etc going to seed – is it likely to be temperature or photoperiodism?*

UNIT 9
Design with microclimate

In this unit students will learn why there are micro-climates and how various factors accentuate or modify climate. They will also learn some strategies to enhance various environments. The concepts raised here will be used again and again in site analysis and design units such as those on orchards, windbreaks and forests.

Learning objectives

Students will be able to:

- Read landscape and predict microclimate effects
- Modify extremes of climate
- Design strategies for small and large landscapes

Terms

Suntrap A relatively still, sun-facing place, sheltered from cold and/or destructive winds and which captures maximum sunlight all day.

Graphics

1. A series of posters, powerpoint slides, or board drawings to show the effect of aspect, slope and solar gain.
2. Draw cold sinks and thermal zones and explain the usefulness of each.
3. Show the effects of vegetation in mopping up heat e.g. Rainforests, mulches.
4. One good drawing of a water body, sea, lake or river to show its cooling effect in summer and warming effect in winter – and its reflection of the suns rays.
5. Show a series of drawings with structures on, that indicate wind funnels, shading, cooling etc.

Ethics

Microclimates can be a source of biodiversity so design to take advantage of them rather than eliminate them.

Principles

- Microclimates give rise to subtle variations in species. For example, one apple cultivar may thrive on heavier, wetter soils than another. This adds to the richness in genetic diversity

- Microclimates can vary over small distances. In a suburban backyard 30 metres wide, one side may be sunny and another shady
- Microclimates need to be designed to maximise diversity – hence flow of energy and cycling of nutrients and pest management

Basic information

Develop a definition of Microclimate by class participation which is something like:

- *Microclimate is the summation of environmental conditions at a particular site as affected by local factors rather than regional ones.*
- *Ask students to do a lot of work on their own places to observe and report on variations in climate, soils and plants in and around where they live. They can then state how they would modify those variations to remove limiting factors.*
- *Ask students how topography (aspect, slope), soil, vegetation, water masses and human structures affect microclimates.*

Topography

The most obvious and important factors affecting microclimates:

Aspect gives rise to different microclimates according to the solar radiation received. Some slopes receive maximum direct morning radiation – with possible frost damage to plants. Slopes which heat up more slowly have less frost damage, and reach a higher temperature. On slopes which are cold and wet, photosynthesis is greatly reduced. These may be good for deciduous fruit.

Slope interacts with cold air, drainage, wind severity. Can use hilltops to intensify wind. Wind speed increases uphill.

Discuss:

- *Solar gain on North and South slopes and shelter from wind (See slides)*
- *Cold sink (See slides)*
- *Thermal zone (See slides)*

Soil

Soil is of lesser importance. However, soil colour, mulch textures and colours, vegetation cover and shade will have local microclimate effects. So will soil deposition from erosive effects of wind, water and steep slope.

Vegetation

Vegetation controls wind flow, and can block it as well. Vegetation can considerably modify climate around human structures i.e. biotecture. Vegetation can create suntraps and a diverse landscape.

Show slides demonstrating effects of vegetation modifying climate. Ask class for a discussion. At this stage talk about what they could do where they live.

The best example of microclimate effects are in small local forests. Forests are warmer on cold days, are cooler on warm days, more humid if it is very dry, and drier if the humdity is high.

Water masses

Large water masses have a big impact on microclimate. They stabilise temperatures. Large masses keep an area warmer, moister and lighter. Even small ponds will increase light by refraction. Temperature can be modified by building a house on top of a tank.

Use slides to show the effect of water masses.

Human structures

These can be effective windbreaks, light absorbers or reflectors. Every aspect of a building has its own microclimate and so has every fence e.g. Shadow, wind, sun, slope etc. Glasshouses and trellises can provide warm humid microclimates and insulate houses. Dark walls reduce frost risk by keeping warmer. Light walls improve ripening by reflecting

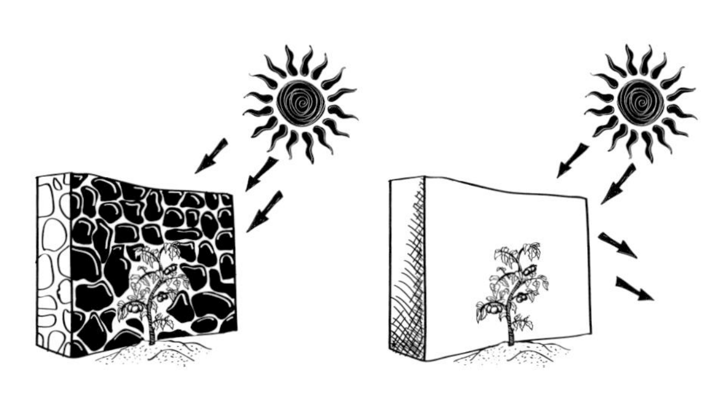

Human structures: Dark walls store warmth, light walls reflect light.

light. North (or South in the Northern hemisphere) facing walls of buildings are very productive, especially in winter. Every house needs a winter morning sun trap. In tropical climates, channel cool winds through buildings.

Use slides to illustrate the major points here. Discuss students' experiences. Mention local factors which can affect microclimate e.g. Tree deflection; temperature on face; weather wear on buildings; animal behaviour under different conditions; human behaviours (e.g. Where people wait for shops to open on cold, clear winter mornings, or where they wait on very hot ones.)

Summer climate control

Student activities

- *Report back next session on soil temperature at different places around a site*
- *Report back on winds, rain and prevailing aspects by looking at wind pruning of plants, and of damage to paint on buildings*
- *Describe which windows have to be shut when it rains. Which doors are shut for hot, and for cold, winds*
- *Ask for designs to modify these climate effects*

UNIT 10
Earthworks

This unit requires familiarity with the units on Soils and Water. This unit is mostly about dams, with some general mention of principles of earthmoving. Concepts of water harvesting and water shed management have already been mentioned. Dams are one way of assisting drought proofing of land. They are especially effective in cool and temperate climates where evaporation is low. Yeomans recommends that 15% of the area of every farm be placed under water. The resultant climate modification and increased productivity will more than compensate. This unit does no more than indicate the pitfalls and possibilities of dams. Students are not experts after having completed this unit.

N.B. In countries where there is a lot of water, this unit must be rewritten to account for systems of ponds and aquaculture. Much of the information in here will remain useful for hilly or mountainous country however. The teacher will need to adapt the information.

Learning objectives

By the end of this unit, students will be able to:

- List necessary consideration for siting dams
- Describe seven types of dams
- Mention important considerations in building dams
- Explain how to remedy leaky dams.

Terms

Core The centre part of a dam wall made of solid clay and well compacted.

Gleying A method of sealing a leaking dam.

Spillway The path on the side of a dam wall which allows floodwater to escape.

Graphics

1. Show again the slides of placing dams on Keylines from the unit on Map Reading.
2. Show the difference between an aquaculture and irrigation dam.
3. Show drawings or photos of several different types of dams.

Site dams to modify harsh climates or, create useful microclimates. Select a specific small area, and consciously put together a number of microclimate factors. Use this area to grow special or out-of-season crops. e.g. Citrus and stone fruit love a morning sun slope above dams. Evergreen fruits to face morning sun above the dam surface.

Ethics

Any earthworks should cause minimal disturbance to soil, water and vegetation, and scars should be healed immediately.

Principles

- Dams should be multipurpose and vegetated
- The greatest amount of water should be collected from the smallest area for the least expense
- Evaporation should be minimised

Two Types of Dam

Irrigation Dams Irrigation dams are high dams, have the smallest possible surface area and are as deep as possible to reduce evaporation. Irrigation dams should feed by gravity.

Aquaculture Dams These dams are low dams, have as large a surface area as possible and don't need to be very deep (2-3m).

Considerations for Siting Dams

Ask students to give the four points and discuss them.

Many countries and states require landowners to obtain permission and/or meet legal requirements before constructing dams, especially dams on waterways.

A poorly sited dam can leak, not harvest water, or in the worst case, collapse.

1. Landform

Ridges, gullies, contour levels, keypoints and saddles are possible sites.

Identify possible dam sites. An ideal place for a dam is just below the keypoint. The most economical site for the dam wall is the narrowest part of the valley.

Contour levels are examined for slope and aspect, which are the next considerations as they are very important for the additional functions of fire control, microclimate and irrigation; eg, given two possible sites (East or West) the dam site to the west will also act as a firebreak.

Gradients are very important because very steep slopes are risky, difficult and expensive.

2. Soil

Test topsoil and parent material. Shale is extremely poor at holding water.

Eliminate all difficult or unsuitable sites. Check the clay by digging a hole 1.0-1.5m deep, or make the ribbon test. Make a clay snake and press flat into a ribbon. If it keeps breaking, there will be too much sand. Compress into a ball, put into water and time how long it takes to disperse.

Visit other properties in the area, check the clay types for their failure and success.

3. Catchment

Type, size, enterprises and vegetation must be thought about. Absorption, nutrient run-off, use and accessibility are also catchment considerations.

Reticulation is the distribution of water from a dam for watering. The angle for irrigation will regulate the rate of water delivery.

4. Ecosystem Cost

Drainage, vegetation loss, habitat loss and downhill disturbance can cost more than the benefits of the water, unless these factors are completely considered. (The real economics!)

Dams and their Functions

Barrage Dams Built across natural waterways, gullies, small streams, even if water is not presently flowing.

Diversion Dams Built where there is no natural catchment. Diversion channels are put around contours to divert water into dams. The overflow from one dam can be channelled along or down to another.

Ring or Turkey Nest Dam For flat lands. Bulldozed out and can be square or round. Often has a windmill.

Contour Dams For valleys and ridges that are not deep or steep, cut back into land and the wall follows the contours of the land.

a. Ridge Dams – A type of diversion dam. They are very expensive. Water is carried to a point and a pond is made.

b. Saddle Dam – Built on the saddle and collects water from both sides.

Perched Dam A small filter dam is constructed above a larger one. This is a dam belonging to the one below. Fish and plant species are separated. The perched dam acts as a silt trap for nutrients or for water cleansing. Both dams can be used for irrigation. Place these downside of roads in valleys to filter the grease etc. from road water.

Bunds These are level banks on graded or flat land, designed to hold water for a specific time. Must be selective where they are used. Primary use is in the tropics to grow rice. They can exist for thousands of years.

Dam Construction

Mark out the dam Find the exact site. Estimate length, width, height, and where the spillway will be. The spillway will dictate water level. Mow around the whole area and don't let grader driver outside this.

Dozer driver Choice is very, very important. It is good to get someone local, as they know the local conditions. Look at other dams they've made and ask people. Find one who will talk to you.

Remove topsoil Be specific. Only remove topsoil up to high water mark. Don't allow a huge scarred landscape. Get topsoil and put in four clear piles. This will cut downwork immensely and cut down loss of topsoil (fewer labour hours).

Construct wall It is very important to remove all sticks, and stones as they will create leaks. Lay the foundation for the wall. Tractor cut out one blade length and build a central core. Use the best clay.

Compacting must be done well. Much more time must be spent in compacting the soil than removing it. Tractors, if they have water-filled wheels, compact better than bulldozers.

The wall should be wide enough to take a car or tractor. See Mollison's books for waterholding formulae.

Stabilise the banks Stabilise the banks with mulch and groundcover. Have hay ready to place over the topsoil to prevent wind or rain erosion. Also broadcast seeds of clover, grains, buckwheat, and plant lemongrass, bamboo and quince.

Don't plant big trees on the wall. They can cause leaks. Good trees around dams are wattles, quinces, bamboos, berry bushes, currants and blueberries. Choose plants that are short and shrubby with fibrous root systems. Avoid things that grow to more than 2m.

Don't establish a full aquaculture (fish species) for 12 months – make this your establishment phase with plants only.

Spillway design The water level is determined by the height of the spillway and the spillway must be at least 0.5m lower than the dam wall. Spillway moves out along the contour and is not directly beside the wall or else water will erode into the wall. Grass the spillway immediately. It is better to first fill with a pipe running to the bottom rather than catchment run-off fill.

Leaking Dams

Clay Put 6" of good clay over the whole dam bottom. This works for many dams which will not immediately hold water but probably will later.

Gleying Clear the bottom of debris and silt. Cover the bottom 10cm with fresh cow or pig manure and/or fresh sloppy lawn clippings. Cover this with carpet, underfelt or newspaper and hay then add a layer of soil and/or clay. Keep it dry for two weeks – then fill the dam.

Bentonite A clay used by potters, bought in fine powder form and has the capacity to hold water by swelling 20x its size. Won't always give 100% success but will dramatically reduce the leakage. The dam is drained, sticks and twigs are removed then bentonite is disc ploughed into the dam.

Terrazole agrosoap/agrosoak A commercial product used to assist in sealing dams.

Pigs/cows Run an electric fence around the water level and let pigs/cows in – hooves and manure have the same effect as gleying.

Ash/lime gypsum Use as a base layer.

Gelignite Used with great success. Broadcast bentonite, lime gypsum, then get a gelignite expert to throw plugs into a full dam. This compacts bentonite (the local farmers co-op will advise on an expert in explosives).

Hessian/cement Dip hessian, underfelt, carpet pieces into a slurry of cement and lay them, overlapping, on the bottom of the dam. Finish the job in one session and don't walk on it or break the surface.

Evaporation from Dams – all dams lose water

Dams are more effective in humid climates than dry climates.

- Cover the water of dams with buoyant styrofoam. Paint it white to reflect heat and stop it breaking down to destroy ozone layer
- Use strategic windbreaks to reduce evaporation
- Have three dams in a row with gravity feeding. When the top one gets low, fill the other two so that evaporation is only happening over two dams instead of three

Irrigation from Dams

Requirement of an irrigation system:

- Water source: dams, bores, soaks, run-off, swales, lakes, tanks and creeks
- Energy to push water: water at head, pressure with pump, solar, gravity etc
- Distribution network: pipes, channels, buckets, net- and-pan, etc
- Emitters: dripline, sprinkler, bucket

Sprinkle Irrigation – wasteful

This is overhead irrigation. A small head of pressure is required.

- This irrigation is inappropriate for highly acid soils
- In arid areas it can cause salination
- The most effective use is usually a well mulched, highly stacked orchard
- Never use overhead irrigation with fungal prone plants
- Use only in early morning and when large water supplies are available
- It loses large amounts of water

Trickle or Drip Irrigation – most efficient

- Best for dryland situations
- It replaces only the water lost by a plant through evapotranspiration
- Plastic piping can be re-used. It is worthwhile investing in a timer
- Always lay from larger pipes to smaller ones
- Good systems for grey water

Flood Irrigation

Used for rice, or as Fukuoka does to weaken one crop so that another can grow. Unless carefully managed, this method can result in salting of soils, especially if fertilisers and herbicides are fed through the system.

N.B.

- Try to use dryland techniques by improving soil organic matter, heavy mulches and deep-rooting species to reduce dependency on watering systems
- It is illegal to put biocides in tanks and hoses connected to town water supplies in case of draw-back
- Irrigation water should be slightly acidic
- Inland water pH is likely to be alkaline
- Don't waste water to running-off from paths, roads, soil etc

Rules for Dry Land Watering

- Irrigate under mulch
- Irrigate at dusk or at night
- Use a timer
- Water for long periods every 3-5 days
- Allow for leaching and replace minerals

Student activities

- *Work out the best watering system for your needs*
- *Return to the map you were given in Map Reading. Locate dam sites and specify which type of dam you would use until you have 15% of the area under water*
- *Do a water audit on the large study and place dams according to physical suitability and enterprise needs*

UNIT 11
Plants in permaculture

Students are encouraged to recognise plants using all their senses – most students revert to the visual sense.

In addition, students study plants as elements in design. This facility is not gained in one lesson and will be complemented in Unit 16 – Zone I and by visits.

Learning objectives

Students will learn to:

- Touch, taste, smell and look at plants when examining them
- Say how they would use plants in Permaculture designs
- List the functions and products of plants
- Carry out simple propagation
- Deduce how climate and micro-climates affect plant behaviour

Terms

Asexual Propagation in which the offspring is a clone of the parents. There is no exchange of genetic material.

Deciduous Plants which lose their leaves at one season of the year.

Evergreen Plants which do not lose their leaves all year and are not bare for any part of it.

Graft To join two plants so that the final plant is better than either of its donors.

Division Dividing plants, usually bulbs, by their new seasons baby bulbs or offsets.

Photosynthesis The process by which green plants use the sun's white light to make starches from water and carbon dioxide. The process yields oxygen.

Propagation All the means used to multiply numbers of plants.

Rootstock The root parts of a plant and some stem used for grafting on to. Generally the rootstock has special qualities.

Scion The top part of a plant grafted on to a rootstock.

Sexual Propagation in which there is an exchange of genetic material.

Graphics

Posters of plants may be useful.

The teaching method for this unit is practical with students examining and naming as many plants as possible and saying what their uses are. Ideally they know where to place them in a permaculture design to strengthen the design or some part of it – or to reduce work or change the climate.

Ethics

Preserve all local regional species and propagate them.

Principles

- Practice thinking of plants as functioning parts in an overall design
- Propagate as many plants as possible for your own use
- Give plants to friends and relatives etc

Explain that climate is a predominant influencing factor in the formation of plant communities.

The elements of climate are constantly changing. Changes are influenced by latitude, altitude, distance from the sea (continental effect), air pressure and air masses as they affect temperature, humidity, rainfall and evaporation.

Soil and landform (aspect and slope) are the next most important factors.

All the above act at the same time but climate is dominant.

In any bioregion there is a large number of species. Local area plants are important for diversity, productivity and survival.

Ask why this is so?
Contrast a monoculture (monospecies) and discuss why this can be vulnerable to many things.

Diversity of planting

- Increases pest predators
- Contains resistant species
- Serves several functions e.g. Windbreak, nitrogen-fixing, productive yields etc

---

- *Ask who knows about PVR. (Plant Varietal Rights). Allow a short discussion*
- *Explain plants can be grouped several ways e.g. According to habit, to products, to soil preference, to seasonality etc*
- *Ask the difference between botany, ecology, landscape design, horticulture and permaculture in how plants are described and used*

Imaginative Ways with Plants – Practical Exercise One

- *Hand out plant specimen to each student. Ask them to describe it any way they like e.g. Poem, drawing. Give about five minutes*
- *Discuss their approaches and findings*

Practical Exercise Two

- *Hand out the second plant specimens to small groups, ask them to describe it – using all senses and presenting their findings any way they like e.g. Words, song, mime, drawings*
- *Ask them to say where and how they would use it in a Permaculture design. 5-10 minutes. Discuss how much richer the information is from a variety of approaches and from a team effort. (Mention that design teams are more creative than individual work)*

Identification of Plants

Initially macro-identification by:

- Sight – for habit, and detail e.g. Tree, creeper and bark, leaf
- Taste – bitter, sweet can identify families, e.g. Mints, eucalypt
- Smell – very important if wanting to remember a plant. Useful for most herbs, and some vegetables e.g. Tomatoes
- Touch – rough, smooth, stinging, slippery etc.
- Hearing – less important but Casuarinas and poplars sough

Then check soil, slope, aspect, microclimate and associations.

Closer Identification

- What is its type? e.g. Tree, shrub, herb etc.
- What are its traits? e.g. Flowers, bark, seeds, leaves
- What does it do (function)? e.g. Mulch, shelter, shade
- What are its yields? e.g. Food, tar, sap, oil, mulch
- How is it propagated? e.g. Seed, cutting, root

Set up the following table on the black/whiteboard and ask students to complete it as a class exercise. Or each group take one heading and report back with its findings to complete the table.

Type	Climber, Shrub, Tree, Herb, Grass, etc
Traits	D or E, Adult Ht, Leaf, Flower, Fruit, etc.
Functions	Mulch, Oxygen, Climate, Habitat, Soil, etc
Yields/Uses	Food, Firewood, Medicine, Dyes, Fibre, etc
Propagation	Seed, Cuttings, Grafting, Division, Budding, etc
Tolerances	Temperature, Soils, pH, Altitude, Pollution, etc

Plant Propagation

- *Ask for the different methods of propagation*
- *Discuss growing from seed, collecting, treating and storing seeds*
- *Explain groups such as stone fruit grafting/ hardwood cuttings*
- *Short discussion on pruning : when and how needed*

Botanical Nomenclature

Explain botanical naming – genus, species and plant families: e.g. Melons, onions, potato, brassica, rootcrops and, especially legumes and their nitrogen fixing capacity.

Student activities

- *Students to learn some new plants by next lesson*
- *Students to bring in extra seed, seedlings, cuttings to share*
- *Bring a list of food plants in students' gardens*

UNIT 12
Forests

Cutting down forests results in drought, water loss, nutrient loss, salted soils and acid rain. These losses are not costed. History has shown that ideologies failing to care for forests carry their own destruction as lethal seeds, e.g. North Africa, Greece, India...

Without trees we cannot inhabit the earth. If we could only understand what trees do for us and how beneficial they are for life on Earth, we would, as many tribes have done, revere all trees as sisters and brothers.

This is a 'keystone' unit. From it students will be able to understand why 35-50% of all land should be under trees. They will know the functional basis of permaculture is a perennial system with an emphasis on trees. Knowledge about forests is basic to designing windbreaks, shelterbelts, suntraps and is necessary for soil improvement, climate modification and optimal growth of plants and animals.

Learning objectives

At the end of this unit students will be able to:

- Define new words that explain the working of a forest
- Explain the impact of forests on climate and soils
- Describe how a young forest functions differently from an old one
- Show how study of forests leads to an understanding of windbreaks

Graphics

Slides simplified from *Earth User's Guide to Permaculture* on whole forest.

Slides explaining photosynthesis/respiration/evapo-transpiration by diagrams. Other diagrams as required.

Terms

Condensation The process by which a substance changes from vapour to liquid state. Generally warm vapour is cooled. Heat is a by-product.

Ekman Spirals A pattern of air turbulence that consists of bands of wind compression on the leeward side of a barrier, such as a windbreak or hill.

Evaporation The process by which a substance changes from liquid to vapour state. The process is activated by heat. It is a cooling process.

Evapotranspiration The transfer of water into the atmosphere through evaporation from the land and transpiration by plants.

Forest A perpetual resource of natural productivity and a basic social resource.

Interception The amount of water held by leaves, stems, animals, bark, webs etc.

Photosynthesis See Unit 11 on Plants in Permaculture.

Respiration The process in plants and animals, whereby O_2 is absorbed and CO_2 emitted.

Throughfall The rain that reaches the ground after passing through a plant canopy. It is rich in nutrients.

Transpiration The process by which plants return water to the atmosphere through the stomata (pores) of their leaves in the form of water vapour.

Trees and water are the two elements required for land repair and for climate change and modification. They are applied as a pattern across the land following the principles of keyline and water harvesting. Most people think of trees and oxygen, however, trees and water are better interlinked and the function of trees with regard to water is more important than that of supplying oxygen.

Equally important is the trees function in tying up carbon dioxide which is now in excess in the air.

Ethics

Save all the forests that remain and plant trees everywhere especially on rain-windward slopes for condensation.

Principles

The function of forests is to give soils the time and the means (humus) to store and clean water, and to catch and store carbon.

- *Write on the board all the words listed in the terms and ask students to form small groups of 3-4 to discuss what they think they mean*
- *Give them 5-8 minutes then ask each group for its meanings. As it finishes, open up discussion to the rest of the class*
- *Have slides ready to explain photosynthesis etc*

- Trees and their environment vary and even defy precise measurement
- A forest is a single assembly, not multiple units.

Forest destruction impacts on soils and water quality

- A young forest does not behave like a mature forest with respect to wind, drought, salt, nutrient uptake and return. e.g. Compare the cost and quantity of food required to feed 10 teenage boys or, 10 very elderly people with no teeth

---

- *Discuss the concept of a Tree Guild or Waru and ask for the fixed and mobile elements*
- *Ask where the Tree begins and ends – slide*
- *Describe parts of the tree – leaf, meristems, bark, roots, stems, twigs, blossoms, flowers, fruit, buds, birds, insects, bacteria, viruses, dust, fungi, spiders, etc*
- *Ask for tree interactions – pollination, pruning, seed dispersal, nutrient uptake, gaseous exchange e.g. Fig bird and figtree cannot live without each other*
- *Use simplified charts to illustrate the following:*

Trees and Effects of Wind

- Deformation
- Tree weight
- Cell specialisation in bark, leaf and bark colour for deflection and for fire resistance
- Size of leaf
- Filtering of wind
- Soil enrichment on the windward side etc
- The great importance of the forest edge and forest closure on the windward side

Trees and Temperature Effects

- Evaporation will cause local heat loss (one medium elm will evaporate/transpire 15,000 lb [6820 kg] on a clear day)
- Condensation causes heat gain locally. By day evaporation cools – by night condensation warms. Air rising over trees lifts, cools and condenses water
- Forests are cooler by day and warmer by night than cleared areas
- Red leaves absorb and reflect heat and decrease temperature
- Damp mulch floor cools a forest

Trees and Precipitation Effects

- Water captured by a forest has a humidifying effect

Trees and Interactions

Use slides and build up a picture while discussing information. Drawings are very effective.

Compression and Turbulence

- 60% of the windstream flows over the forest by deflection
- 40% of the windstream flows through
- At 1km inside a forest it is still
- Dry hot air entering the forest is cooled, humidified and stilled
- Cold, wet air entering the forest is warmed, dried and stilled

Deflected wind

- Deflected wind rises above the forest and is compressed and effective to 20 times the height of the treeline, it cools the air and causes rain

- Rain from the sea has an iodine molecule as its nucleus, forest seeded rain has forest dust particles, viruses etc
- The lee side is drier and less nutrient rich but Ekman spirals affect local rain leeward from the forest

Discuss rain nuclei.

Condensation Phenomena

- Condensation occurs on sea-facing coasts and islands. Often there is no rain only condensation, as in a drip forest. A single tree can present 16ha of laminate surface for condensation
- Fog has higher precipitation than clear air. Not measured in rain gauges

Rehumidfying Airstreams

- Oceans give water and later rain on forests
- Forests return 75% of their water to the air
- 60% of inland water is from forests
- 90% of water absorbed by roots is lost through leaves

From Cloud to River

Using slides to discuss rain over whole tree, dripline, bark absorption, cleaning, nutrient collection.

- In forests, snow is held on trees or on the ground under trees where it melts slowly and is absorbed into the soil.
- In deforested areas it melts much more quickly and either causes floods or sublimes (evaporates) – either way the soil does not benefit

Student activities

- *Stand in a forest when it is raining and verify interception and throughfall*
- *Sketch the pattern of throughfall.*

Extra notes

In Cambodia, when people have built a wooden house and are moving into it they have a ceremony with the monks from the pagoda to placate the spirit of the Forest for having used the timber.

UNIT 13 Windbreaks

Windbreaks are basic to permaculture design and function in many ways. Windbreaks carry out many of the same functions as forests and can be seen as an extrapolation of them.

Learning objectives

By the end of this unit students will be able to:

- Explain how windbreaks work
- Explain why windbreaks are necessary
- Design windbreaks which are site specific i.e. Suit that piece of land
- Design windbreaks for specific purposes, e.g. Firebreaks, shelter belts, suntraps, privacy, food forests, orchard protection

Graphics

1. Drawings of the height and shape and protection given by a windbreak.
2. Sketches of five types of windbreaks e.g. Alternate, Compound, Incrop etc.
3. Sketch of a network of windbreaks for a village orchard in cyclone areas.
4. Drawing of a garden windbreak e.g. One metre high and temporary – any photos or illustrations of windbreaks would be useful.

Terms

Chill factor The additional lowering of temperature caused by wind moving over plants and animal species removing heat and moisture, thereby reducing the actual thermometer temperature.

Compound One of two types of leaf description. A compound leaf is one with many small leaflets like a jacaranda.

Cyclone This is an area of low pressure which in tropical regions can be of great force and very destructive.

Fibrous Having a rough texture. Usually refers to bark which, being rough, absorbs wind.

Parabolic or luneate Decribes the structure of windbreaks so they work to lift wind up and over a cultivated area and deflect horizontal winds around it.

Permeability The ability of soils or rock to absorb water.

Pioneers The first plants to grow on bare ground after it has been destroyed by fire, clearing, sprays etc. They condition the soil for the next flush of species.

Windbreaks are important on every site since they serve so many functions. Apart from soil protection, improving water infiltration and giving products, they can increase productivity by protecting the cropping fields. However, in some areas it is possible to use trees to the opposite effect from windbreaks by pruning off all the bottom branches and increasing the speed of the wind under the canopy. This is done in very hot places to create a low, cool ground wind to cool buildings and animals. In other cases, windbreaks are designed to speed up the wind which is being used to generate wind energy systems.

Ethics

Windbreaks are a substantial part of the 35% tree cover that is considered minimal for any farm or landscape.

Principles

- Sites have predictable wind patterns which are obtainable from records, windpruning and building wear
- Design buildings, garden and animal shelter to face the sun in cool climates. In a hot climate they should be shaded
- Windbreaks can work as suntraps, shelter belts, wind funnels or wind filters

Revise basic concepts from Climate – Ask how winds work. (see below)

Cold winds remove heat from surfaces, plants, buildings, water and the living bodies of stock and humans. With wind velocity and evaporation of fluids a chill factor is created. This is good in a very hot climate but if the wind is already very hot, then plant and animal growth stops and even dies.

Cold winds results in:

- Cooler climates than temperatures show
- Retarded plant growth
- Height and yield of plants decrease
- Solar devices and insulation work less efficiently

In cyclone areas, winds are decisive.

Disadvantages of Winds

Ask students to give their ideas.

Houses

In winter, cold winds remove 60% of heat via windows and the solar hot water system collector. In summer hot winds create uncomfortable living environments leading to the use of air conditioners and increased electricity usage.

Stock

In winter, animals lose 30% of body weight in three days in cold winds:

- 20% of Australian sheep losses follow wind chill warnings
- 16% less is eaten by cattle on exposed sites

In summer, during heat waves animals have a reduced food intake.

Household windbreak in cyclone region.

Erosion

Without windbreaks, soils lose more moisture, dams silt up and evaporate faster and rivers silt up and flood more often.

Plants

In summer individual plants stop photosynthesising and wilt when there are hot dry winds.

In winter, cold winds can kill them.

Orchard yields can be destroyed at flowering by frost, hail, very hot or very cold winds. Field crops can be destroyed at any time by gale force winds.

Advantages of Winds

Advantages of winds? Ask students for answers.

- Light winds remove humidity and help plants to resist fungal diseases
- Light winds cool houses and people
- Some winds carry warmth and moisture
- Winds can be used to generate energy
- Winds can carry seed and pollen
- Wind aerates water systems such as rivers and ponds
- Winds help to clear polluted air from the cities

Ask students for advantages – remind them that some of these are the opposite of disadvantages of strong winds.

Practical Advantages of Windbreaks

Houses

Save 30% of heating fuels in even moderate climates. Save 15^{0}C at ground temperature with windbreaks.

Animals

It is considered that no part of a paddock should be more than 400m from good shade, otherwise losses in moisture and energy from stressed animals trying to keep cool are greater than gains in meat/wool/ milk production.

Erosion

Soil losses can be reduced 50-70%. Dust has settled completely at 100 metres into a tree belt. Dams and ponds have reduced silting and evaporation of water.

Plants

Frosts are reduced. The effect is up to 10 times the height of the windbreak. Overall, orchards have an increased net production of up to 25%.

- Damage is reduced by 50% in citrus orchards
- Blossom set and seed set are increased
- Also there will be increased pollination
- Branch breakage and uprooting is decreased

Civil Construction

- Windgaps or funnels increase wind velocity and cooling
- Windbreaks prevent snowdrift and contain the snow melt
- Windbreaks protect roads and caravan parks
- They assist human health around settlements in arid areas and polluted cities

Windbreak Designs

- Up to 35% of the total site area to be planted as a succession
- 50% permeability = 50% windspeed reduction.
- Windbreaks are effective to a distance of 27 times the windbreak height, if permeability is 50%. i.e. see Unit 12 on Forests and how they work
- Height and density is critical to permeability.
- Fine, dense foliage is required to remove heat and dust
- Bare stems for cooling, these have a thick evergreen canopy to cast solid shade
- The edge should be permanent and at an angle of 45^{0} to lift the wind up and over the paddocks.
- Diagrams of a windbreak both in plan view and in elevation

Designs

Traditional straight-line windbreaks create eddies of violent winds over fields. A curved or boomerang shape, called parabolic, will lift wind over crops, and will direct ground wind away.

Drawings of all of the following Windbreak types:

Alternate used in coastal areas to reduce salt burn: Fine leaf, Zig-zag, 5-row, Westringia, Araucaria, Casuarina

Compound used in deserts to reduce dust: High density, Mounds, Grass, Herbs, Shrubs, Trees

Permeable used in Inland Frost areas to create air movement: Standard shrubs, Air movement, Move frosts

Incrop for Tropical Sub-tropical cyclone areas: Absorb wind (cyclone), Mix fruits with trees, Filtered light.

Parabolas for Cold Climates:
Evergreen, Leaves to grounds, Wind lifted up and around, Straight species on lee, Permanent edge

Materials Trellis, brick, earth mounds and tyres in a parabolic or luneate shape to deflect the wind sideways.

Windbreak Design of Orchards and Cropping Fields

Networks 10-16m high. Orchards should be planted out into networks of windbreaks which are 10-16m high and 33-66m wide. Width depends on final adult height of orchard species, 33-66m wide.

Garden Windbreaks Temporary summer windbreaks are effective, e.g. Corn, sunflowers, Jerusalem artichokes.

Wind Sensitive Plants Citrus, avocado, kiwi fruit, deciduous fruit, corn, sugar cane, bananas.

Windbreak Essentials

Pioneers Provide initial protection, water, anchorage and sometimes mulch.

Fire & Wind Resistance Fibrous stems, Fleshy leaves, Needle-like leaves, Furry leaves, 40-50% penetrability of front line or dominant species for fire resistance.

Early Fast Growth This is all important. Use pioneer species. Fences 18-24" high will grow potatoes, strawberries and cabbages behind them.

Protection Open ended plastic bags, Earth Mounds, Brush Fences, Tall grasses (e.g Banana grass – bamboo in Alice Springs) mow later.

Species Choice Nitrogen fixers, Heavy leaf fall-mulch potential, Forage plants, Barriers to stock and weeds, Bee forage, Timber, Fuel, Home to predatory insects, Wildlife corridors, Good species are: Acacias, Casuarinas, Tagasaste, Leucaena.

Consider important factors such as whether evergreen or deciduous, mature height of trees, branch drop, leaf size and plant habit.

Student activities

- *Ask students to design a windbreak for a site they know*
- *Ask for a very clear site analysis first and they must show how their design will correct the wind, shade, erosion and other factors*

SECTION 4

Designing productive ecosystems

UNIT 14
Patterns in nature

This is the most philosophical unit in the whole course. It can sound like magic or something more abstract like chaos theory, or even nuclear astronomy, which it resembles.

Some students find the concepts and implications difficult, so I teach simultaneously at two levels. For one group of students I am content if they understand that there are principles behind patterns and their connections and functions in permaculture. For others I indicate the depth behind the subject and lead them to further reading. Most patterns are integrative.

In permaculture:

- The pattern is the design
- Design is the subject of permaculture
- Designs are compiled from an assembly of components or forms

Patterns have several parts. Firstly there is a form. This form is repeated. For example one hill exists as a form but repeated across the landscape is a pattern of hills with particular aspects. Recurrence is basic to a pattern. Patterns normally have symmetry, and balance. There is an array of forms in nature. Several forms and patterns can interplay such as the branching pattern of a tree canopy and the circular stems of the leaves and branches. Different types of symmetry are important. A butterfly needs symmetry and so does the canopy for a tree for its leaves to access the sun. There are several forms of symmetry.

Patterns in nature have functions. We use forms and patterns to achieve special design functions. This is explained further in this unit.

Learning objectives

My learning objectives differ according to the ability of the students, however by the end of the unit I hope they will have learned to:

- See patterns that already exist
- Apply zone and sector planning as pattern application
- Design patterns with understanding of their impacts
- Impose a design as an assembly of components

Terms

Deflect Turned aside, bent to one side.

Edge effect Or ecotone, is the transitional area between two ecosystems and is effectively an intergrade.

Ekman effects Is a mathematical construct showing the vertical profile of vectors representing wind strength and direction at various heights. Also called Ekman spirals.

Flow Is the movement of fluids such as wind and water, and has wave forms.

Harmonics Are the maths applied to the relation of quantities whose reciprocals are in arithmetic progression (e.g. 1, 1/2, 1/3, 1/4) or to such points, lines, functions involving such relations. They are also understood as waves and fractils.

Orders Changes in size or scale; measure of size, volume or concentration contained within the same form or pattern.

Principles

Aim to:

- See patterns that already exist
- Impose patterns to achieve specific results
- Use edge effects and harmonics (waves/vibration/repeatability)
- Manipulate flow of air and water

Role of designers:

- Keep the design small
- Keep it varied
- 'Stupidity is an attempt to iron out all differences and not to value them or to use them creatively' (Mollison)
- Introduce the concept of time as a pattern, and the world as a sequence of events within a pattern. Some time patterns are relatively predictable e.g. Seasons
- Explore how many time patterns from small to large students can find e.g. With climate patterns we can harvest and make designs for different harvest dates. What else is possible?

Symbols

Symbols are derived from form and patterns in nature and come to represent concepts in peoples' minds.

Ask students to name a few you draw and ask them how they respond to these symbols. e.g. Swastika, footprint, yin/yang, mandalas.

Look for relationships among patterns and how they mould each other. For example, song patterns: How long does it take people to learn a new ballad melody with its repeatable verses? Discuss microclimate interactions of the elements such as water on site and soil, and so on.

Patterns have orders and flow

Orders are the measure of size, volume or concentration contained within a pattern. Orders are related to gathering and dispersing of contents e.g. Nutrients, waste and energy flows. There is slight energy in small waves and great energy in large ones. Orders are numbered from one to a maximum of seven but they may be as few as five.

- The first order 1 is substantial, sluggish and has inertia
- The last order, 6 or 7, is slight, fast, turbulent.

Look at, or draw a tree and show that the trunk is order 1 while the new leaves at the tips are order 7. Each branching leads to a new order and reflects a growth pulse.

Discuss whether any order is more important than another. Don't confuse stability of orders with status and power, as though the tree stem were less important than the leaves or vice versa. N.B Saying – Out of order. Orders are discrete and anything somewhere in between will either shrink back to the lower order, or increase to the higher order. Otherwise they are out of order.

From the concept of order, in permaculture we refute the concept of status and assert that of function; that it is not what you are, it is what you do, in relation to the society you choose to live in. We need each other and it is a reciprocal need wherever we have a function in relation to each other.

Ask students to give more examples of the difference between status and function. In permaculture we work with the function not status or purpose.

In applying designs, students need to understand which order of landscape they are working in. For example, if they are at the top of the watershed, then they will be primarily working with Order 6. If they try to apply an Order 1 solution such as a huge dam – it will be out of order and probably fail. If they are in the inner city the order of tree they design for sites needs to be Order 5 or 6. For example, don't plant giant forest trees in inner city backyards. Much poor design work is because of lack of understanding of the orders in the landscape.

Flow is movement over time and space and flow can be deflected, or impeded. Some forms are specially adapted to deal with flow e.g. Circles. Others draw flow to or through them e.g. Curves in rivers, canyon effects of winds around city buildings. For flow to occur there must be access. This is important in thinking about flooding and for winds – it assists in dispersing energy.

Understanding different patterns

Linear

Straight lines are weak forms e.g. Boxes, triangles. They are easy to break edges or corners, and are easily invaded such as rectangular garden beds.
Advantage: good for direction finding such as road maps.
Disadvantage: hard to remember.

Non-linear

Usually stronger forms such as songs, ballads, dances and are more easily remembered and derived from nature.

We spend the rest of the unit on these.

Non-linear patterns include branching, networks, circles, spirals, tessellation, luneate as well as irregular patterns.

- All these forms have special functions
- All patterns also have mathematical bases
- Many patterns also apply to human societies

Network patterns have been most recently described as the science of networks. They are almost certainly the most important in permaculture. All feeding relationships, social relationships and nutrient cycles are networks. They consist of two elements – links and nodes. They are very strong if they have many nodes.

The number of links to a node reflects its necessity. But links can also connect to many nodes. Some examples are pollinators such as bees which are links and visit many species and pollinate about a third of the world's foods. If there is only one node i.e. apple tree in a bare field, it is very weak. The greater the number of species visited by bees, the stronger the network. Nitrogen fixing trees operate as strong nodes in soil ecology. Mycelium is the best example of a network delivering and receiving materials and breaking them down in permanent species landscapes. Networks are hard to destroy because they are non hierarchical. It is possible to remove many links and nodes and yet not destroy the whole structure. They are important in social peoples' organisations which want to perpetuate themselves. They can break into smaller parts if they get too big.

Branching patterns comprise those forms which resemble trees and rivers. They have a central trunk and then branch and branch again. They have up to five orders. Show a graphic or draw one.

Ask class for examples of branching patterns. They should be able to think of at least ten.

Orders in Trees

From trunk through to new leaf tips – 5 orders.

Orders in Human Settlement

Order 5 – Hamlets 20/200 people
Order 4 – Villages, about 1000 people
Order 3 – Towns, from 5000-30,000 people
Order 2 – Cities, from 70-100,000 people
Order 1 – Megopoli, millions of people.

The branching pattern delivers and returns substances, e.g. Energy, water, nutrients, messages from one order to another. Water, for example, is taken from the earth and delivered to the leaves. The branching pattern facilitates this. What is being delivered or returned in other orders students can think of?

As the order increases, flow slows down. All warm-blooded animals have approximately the same number of heartbeats during their lifetimes. There is a correlation between volume and velocity – the speed of flow. Consider traffic as it moves from city arteries to country towns. Most things that branch and flow have about six orders. The largest order is very slow.

The mathematics of branching for each order are:

Order 1 1
Order 2 1 +1/2
Order 3 1 +1/2 +1/4
Order 4 1 +1/2 + 1/4 + 1/8
Order 5 1 +1/2 + 1/4 + 1/8 + 1/16

The total of the orders can never reach 2. The volume of the largest one will never be exceeded by the total amount of the smaller ones.

When you look at the orders as a pattern you can see that they have similar properties. For example, order 1 of most rivers is estuaries and these are more similar than an entire river system which will be between orders 1 and say, 6.

Orders in rivers show that order 1 in different catchments have similar ecosystems and so do the other orders.

Orders in veins and arteries show the same patterns e.g. Order 6 veins in the fingers and toes are more like each other than they are to the arteries. This has implications for design and function.

Draw a branching system and show orders, link up pulse points of growth, also known as events, and mathematical relationships of 1, 1/2, 1/4, 1/8 i.e. never reaches two. Also the Order 2 is half the length and half the diameter of the Order 1.

Application of branching patterns in design: Delivering reticulated grey water to fruit trees. Ask for others.

Circle patterns are strong patterns. Circles have the smallest perimeter for their area.

This is expressed as 'πr^2'. With their smaller perimeter they contain their contents very efficiently and having no straight edges they are less likely to be damaged or invaded.

Examples: Stems, vessels, cross-sections, eyes, horizon, architecture, Blackfeet Indian and Maori cultures.

Ask for 20 examples of circles from nature and ask why, in each case the circle performs the best function. Show the circle poster from the Design in Nature series and ask why eyes and fruit are round.

Application of circle patterns in design: Where resources are few, or need to be contained as in gardens in arid zones, banana circle for grey water, water tanks and irrigation ponds and lakes.

Spiral patterns conserve strength and are protective. They are based on the Fibonacci series. 1+2 = 3 + 2 = 5 + 3 = 8 + 5 = 13 + 8 = 21 + 13...

Ask for examples such as wind, shells, waves, willy-willies (dust devil), water flow (currents), plant nutrients moving up tree trunks, herb spirals.

Winds and their orders

Order 4 – Willy-willies, tiny spirals, very fast, over flat hot surface and last a few minutes
Order 3 – Tornados, large downward spirals and last several hours
Order 2 – Hurricanes/cyclones. Huge. Last 3-6 weeks

Order 1 – Planetary frontal systems. Last for months, cause monsoon climates.

The Kopago Indians of North America used their moon spiral knowledge to forecast changes in intensity of life e.g. Cyclones, floods. The eleven-year moon cycle can be related to a 3% change in the intensity of life and from this sunspots, earthquakes and volcanic activity can be predicted.

Application of spiral pattern in design: Where space is reduced such as herb spirals, and where wind is strong such as desert area tree seeding for regeneration.

Tessellation patterns are shapes made to fit any surface. In their various forms they are elongated, compressed etc. Many body forms are lobular and tessellated shapes consider shoulders and kidneys, also tennis balls, Earth's tectonic plates, bones, turtle shell, barks and dried mud.

Application of tessellated designs: Stone mulches in dry areas. Straw mulching over irregular land.

Luneate or serpentine patterns are those which are crescent shaped. They usually have a stronger side and a protected one. Some examples are: Dunes, lakes, moon at different times, snake movement, erosion, songs, contour lines and windbreaks.

Sand dunes and their orders

Order 1 – Draas, great dunes; no perceptible movement in hundreds of years.
Order 2 – Zourghs, starforms; few metres every hundred years.
Order 3 – Dunes, with several forms all having a low long back in direction of prevailing wind steep face on leeward side.
Order 4 – Ripples are small dunes moving several metres per month.
Order 5 – Saltation is when 50% of sand moves often several metres per day and leaves ripple marks as wave 'harmonics' can
Application of luneate patterns: windbreaks, mounds, flood water capture from rivers.

Irregular patterns

Scattered particles such as dust in the air, lichen on rock, Section of monocot stems etc.

Streamlines are like horizontal streaks in wind, dyes in water, clouds, plane lines.

Applications of irregular scattered patterns: broadcasting seed, pellets, shade over a site.

Disintegrative patterns provide the materials for the integrative patterns, for example, seed scatter, erosion, flowering. The volcano Mt. Helena which erupted in USA is now highly productive with its rich basaltic soils.

Applications of disintegrative patterns: floods across plains, compost into humus, landslides.

Applying pattern knowledge

Wind flow and spirals Windbreaks, rivers, water-holes, deliberate controlled flooding, planting in curves and circles, water harvesting from rivers.

Time cycles affect yields, old sayings, observation of correlated events e.g. When corn is knee high plant the beans with it.

Social effects Sizes of towns, types of occupations; these can succeed or fail due to imposed patterns of occupations and transport lines.

Edge effects/ecotones Humans are an edge species and most live beside the oceans or rivers. There are also implications for alley cropping and grazing and these have an impact on harvesting and design harmonics.

Student activities

- *Look for patterns such as weeds along roadsides*
- *Discuss patterns in and around their own house and garden Show how patterns can be imposed on existing site to:*
  - *Decrease work*
  - *Increase yields*

UNIT 15
ZONE 0: Siting, building and furnishing eco-homes

Low energy houses often look conventional and are beautiful. They give a comfortable environment while saving on heating, cooling and water bills. They require the resident to act and manage energy flows such as closing curtains at night to hold warmth in the room.

Buildings consume resources. Most new buildings including homes qualify as one or more of the following:

- Consumer junkies of huge quantities of non-renewable resources
- Toxic in terms of glues, paints, and furnishings etc. and for energy, air, water and waste
- Vulnerable and meet none of the owners' needs for energy, food and water

Because buildings are voracious consumers of resources it is valuable for students to have an understanding of the nature of resources. Their ability to be re-absorbed into ecosystems can be seen as a continuum. For example:

- Uranium (10,000 years to cycle)
- Gold and coal (a long time)
- Timber (one to many years)
- Straw (a few weeks)
- Nitrogen (can be 24 hours)
- Wind (instantaneous)

When building we want to choose resources which will recycle perfectly. And that is none of the plastic products. We need either those which last a very long time with little maintenance or, those which compost perfectly such as leaf and bamboo houses. At each step in building design we consider carefully for every item, the resource and its final state.

This unit explores the basic principles of designing and building a home:

- Siting
- Orientation
- Layout
  - · Human behaviour and needs
  - · Appropriate technology
- Construction materials
- House furniture.
- Houses for arid climates and hot wet climates

Learning objectives

By the end of this unit, students will be able to:

- List factors for consideration in siting a house
- Orient a house correctly for different climates
- Design a house based on human behaviour, energy saving
- Choose appropriate construction materials
- Select good insulation
- Suggest environmentally sound house fittings

Teaching tools

- Drawings of different types of traditional houses which show the best features of energy saving and local materials
- Examples of house siting and orientation for different climates and slopes. Show where the water resources are and where the grey water will go
- Collect different types of insulating materials

Terms

Active designs Where energy inputs from outside and usually non- renewable, are required.

Biogas Gases generated from waste such as animal manures. It burns cleanly. It is an efficient way of recycling waste.

Passive designs Those which need some adjustment e.g. Heaters or curtains at night.

Reactive designs Those sophisticated designs which adjust themselves to changing conditions.

Retrofit A word describing the taking of a conventionally built home and making it more energy and water (resource) efficient.

Thermal mass A solid body of material used in buildings as floors, walls etc. to absorb heat and later radiate it back into a cooler space. It acts as a heat store.

Discuss the 'sick house' syndrome. Glues, paints, cleaning chemicals, polyurethanes.

Ethics

All structures should be environmental living zones built to preserve non-renewable resources and create zero waste, use few external inputs and last a long time and be beautiful and comfortable.

Ask students if they can name four principles which would enable the ethics.

Home powered by renewable energies

Principles

- Admit and store sun's energy when needed, and remove and exclude heat energy when not needed
- Generate designs which reflects simplicity, economy and resources recycling
- Design to related space and function
- Design or retrofit your house yourself
- Design and build houses which meet their own needs for renewable energy, food, process waste, use only natural light during the day

Siting houses

Houses are best sited on keypoints on sun facing slopes with land behind for orchards, and with high dams in forests.

Show drawing in transverse section. Ask students what factors to consider when siting a house.

Factors	Indicators
Access and position	Erosions, escape path in time of disasters
Climate	Winds, floods, pollution drought, frost, reliability, predictability and possible disasters
Soils	Drainage, suitability for foundations, or building materials
Surrounding landuse	Traffic, development proposals, industry, toxins, history of land
Topography	Aspects – thermal belts, solar gain, shading, Slope – contours, keylines, keypoints, flood plains
Water sources	Springs, tanks, dams, creeks, recycling water, solar HWS placement, precipitation
Vegetation	Valuable natural vegetation, time to restoring vegetation, biodiversity

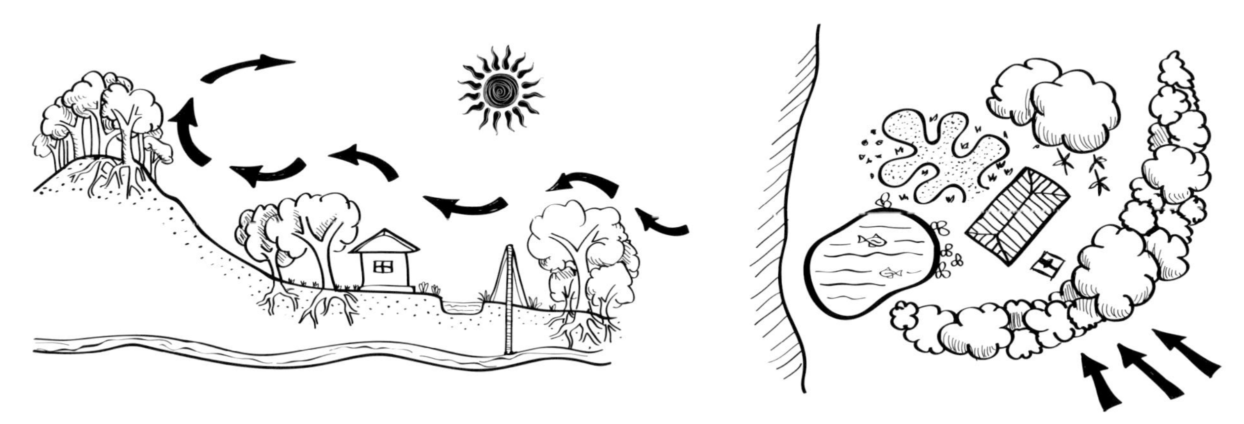

Siting house with micro climate

Buildings and utilities around the house

- Place utility buildings such as store rooms, animal housing and water tanks behind the house
- Obtain drinking and cooking water from high rain tanks, clear springs or high deep dams behind the buildings and carry water downhill and gravity feed it to buildings
- Place water tanks close to houses for appropriate temperature control and insulation: tank on cool side in hot climates and, on north-west or south-west as appropriate in cold climates
- Use creeks, ponds or dam water for washing, laundry and recycle to the garden while functioning as fire control. Use gravity even if there is less pressure
- Use dam wall as a roadway to consolidate it and prevent potential erosion. Select vegetation as deciduous or evergreen, to cool or permit winter warming

Never build on a south facing slope in cool southern hemisphere. Corollary, do not build on a north facing slope in cool northern hemisphere.

Students now look at their drawing of their home, or one they know very well. They now assess it in line with the factors just mentioned.

Orientation

This is most important in designing passive solar houses. It must be placed on the land so that it gains maximum solar gain in winter and can expel or avoid heat in summer.

The further from the Equator the more the house and its roof must be angled to maximise the sun's energy. Show drawings.

So orientation depends on latitude and the following is based on Sydney. Tables and websites are readily available to calculate the angle for all latitudes.

- House should be on an east-west axis, 18^{0} east of north which gives good morning and all day sun without the need for daytime artificial light and excellent winter warming in the southern hemisphere
- Roof Slope – 31^{0} roof: ceiling. This gives maximum solar gain for solar power and heating, and skylights
- The angle of the eaves should be 64^{0} from sill to eaves, or the eaves should shield the glazing from summer sun and allow winter sun to enter

---

Students to place house in drawing so that it will have as even as possible temperature all year, or they can heat it or cool it from natural renewable energies.

House design

This is how the rooms function according to their use to achieve

- Summer cooling
- Winter warmth

The north or south facing aspect (depends on which hemisphere) on an east-west axis does this most efficiently.

Houses acquire value where there are outdoor and indoor recreation areas. Shade needs to be carefully considered for comfort during different seasons and times of the day.

Room functions Position active, most used rooms to the sun where they are naturally lit and warmed. Kitchens on sun-east corners are good for getting the sun at breakfast. Offices can be on the west. Bedrooms and bathrooms which are used less often can then be to the south or south-east in the southern hemisphere and north or north-east in the northern hemisphere. It is pleasant to wake to morning sun.

Hot climates need covered verandahs to the west. Open sunny pergolas with deciduous vines will control heat during summer in Mediterranean climates.

Technology Use a considerable amount of glass to the sun with some thermal mass on the wall to store heat. Floors of brick, tile, stone, concrete or dirt store heat which radiates back into rooms in the evening. West walls can be designed to hold heat if required, and glazing should be reduced to the south – non-solar aspect. Always have fireplaces on internal walls. Some low technology heat stores are glass bottles of water or sump oil, or, trombe walls.

A Trombe wall is a concrete, mud or rammed earth wall, painted a dark colour and facing the sun. Glass on the outside forms a cavity, and vents at top and bottom draw cold air from the floor into this cavity. As the air heats, it rises and moves into the room as warm air.

Students now design and retrofit their homes according to the information they now have.

Houses in different biomes: (a) Humid high altitude farm and house; (b) Coastal cyclone farm and house; (c) Humid delta farm and house.

Construction materials

Choose local resources, products and materials and those which suit local conditions. Brick walls are more efficient than plaster. The ideal house is an inside-out brick veneer. Hot water pipes from convection stoves can run under floors. Vents are placed to allow and exclude heat from entering other rooms.

Insulation is important as it keeps homes at an even temperature i.e. winter warmth in, and summer heat out.

House of stone, mud, wood and thatch: all local materials

Electricity efficiency Recent research demonstrates that small new technologies can make huge savings in homes and for nations. Such things as the compact fluoros light globes which cut energy use by 80% save coal energy.

Simple efficiency in refrigerators, washing machines (recycle water) and in stoves, means great national savings.

Some facts about houses and energy. In Australia 20% of all energy is used in homes. The rest is consumed by industry and transport.

	Total kilowatt hours	Hot water	Space heating
Subtropical	8,000	50%	5%
Cool Mediterranean/ temperate	14,000	30%	50%

Irrespective of climate, hot water is one of the biggest consumers of energy, so solar hot water heaters should be used universally, especially with homes with more than one person. A two-car home uses the equivalent of 40,000kw per hour year.

Space heating principles aim to:

- Admit renewable energy sources
- Hold them with minimal loss

Energy losses from a warm room Once you have paid/supplied energy to warm a room, you want to reduce losses to an absolute minimum.

Floor 15%
Ceilings 30%
Windows 25%
Walls 20%
Other 10%

The above figures show you when to start to make savings.

Insulation is critically important in keeping excessive heat out and holding reasonable heat in. That is, helping to maintain a comfortable temperature all year around. There are different types of insulation.

Discuss the pros and cons of the following:

- *Ceilings: Consider fibreglass, cellufoam, insulwool, newspaper, Stramit straw, seagrass*
- *Floors: Insulate from below with ceiling materials if wood floors. For concrete slabs only 1m around perimeter is necessary. Use carpet and underfelt in inside rooms with overcast climates or no solar gain*
- *Walls: Use brick, mud, stone, unpainted*
- *Windows: Double glazed, properly sealed in winter. Use pelmets on all lined curtains or hot air will escape*
- *Roofs: The best is a sod roof with 4" of dirt. Galvanised roofs must be insulated*
- *Other: Seal door frames, close off chimneys etc*

Ventilation is necessary to cool houses and to keep them dry in humid climates. Solar chimney: Painted black to draw up hot air which accumulates in the ceiling.

Vents: These in house walls beside a shade house allow cool air to enter in summer and can be shut in winter.

Cool air tunnels: A tunnel from outside to inside the house at 18" depth reduce room temperature. They must be sealed against insects and pests.

Students discuss with each other in pairs what building materials they could use which are renewable, local and recyclable.

Biotecture is using living materials to modify climate.

- Evergreen pergolas and vines on the west protect walls against hot sun
- Evergreen vines on the south (cool) side assists insulation effectively
- Tall trees on the north (solar side) with a slender trunk allow winter sun through and shades the roof in summer
- Deciduous trees on the north (solar side) and west, planted on an earth mound, will allow the winter sun and give summer shade
- A shade house on the south provides extra insulation in cool climates, and, a cool retreat place in hot areas
- Consider plants for suntraps, windbreaks, trellises, fences of stone, earth, plants, weaves, rails etc

---

Ask how these could be used to modify climates.

House furniture Whether made or bought they should have these qualities:

- Energy efficient: use renewable resources
- Durable and repairable and recyclable
- Non-polluting
- Resource conserving
- Local
- Efficient with maximum effect for least input

Wet areas Bathrooms can be used as glasshouses on the north-east corner (Southern Hemisphere) and water recycled immediately to the garden, and heat to the kitchen.

Hand basins can be connected to the toilet cistern.

Kitchen and bathroom plumbing can be connected to carry hot water from the slow combustion stove.

Kitchens work well if they open to the garden. They can share heat or stove with a living area. A hot water sleeve on the stove can fill a roof tank and gravity feed back to the kitchen sink and bathroom.

Stoves Lorena stove in summer, some cooking can be done outside on this.

A radiant heat stove, Kuchelofen is the most efficient.

Bakers ovens, rocket stoves and pizza ovens, can be used for groups or communities.

Hay Boxes, also known as fireless cookers, are very efficient and have been around for a long time.

Solar stoves which concentrate sun rays are now possible especially for arid and warm areas.

Producing biogas

Biogas, which is methane, is available from rubbish dumps, or can be made from the digested manures of domestic animals. It burns very cleanly to CO_2.

Heating and hot water Hot water systems can easily be made from batch heaters, brick boxes, solar heaters, Jean Pain Method and solar ponds and gas bottles in Alice Springs.

Energy sources to consider: Hydro, hydraulic, wind, geo-thermal, gravity, timber.

Heaters There are many types of heaters for space warming.

1. Open fireplaces burning wood or coal – at most 25% efficient – and potbelly stoves are terrible.
2. Airtight stoves burning wood or coal are 60% efficient.
3. LPG or natural gas with flue is 75% efficient.
4. However, bottled LPG cost nearly 2x natural gas and bad for environment.
5. Electricity is 100% efficient but more expensive than natural gas which often uses coal Bar radiators and fan heaters are very expensive and inefficient.
6. Most expensive is bagged wood in open fire – $175/tonne. Cheapest is airtight, slow combustion stove with bulkwood at 0.98 cents/mJ. ($175=8.02 cents/mJ) (1990) Next is natural gas in portable heater 0.99 cents/mJ.

Sewage and water Rainwater coming off farm buildings can be gravity fed to the house. This can be used for cooking and drinking. Other water can come from dams, creeks etc.

Fit showers and taps with low-volume delivery and use 1/2-flush mode toilets.

All grey water should be recycled and all house water can be fed out to either the food gardens or timber forests.

Some compost toilets are: Biopan, Envirocycle and Clivus Multram Nature loo all dispose of sewerage on site.

Ask students what they know about different sewerage systems. Discuss council regulations, then get students to discuss again in pairs how they will heat, cool, furnish, cook in their houses with sustainable resources.

House design for different climates

Temperate Climates

Classic pattern for solar house used in this unit.
Celestory windows allow sun to enter to rear rooms.
Maximise ventilation
Pergolas
Insulation
High thermal mass
Fans
S-shadehouse
Heat banks
Draft-proof

Dry Heat Houses
Underground, or nearly
Wide verandahs, shadehouses
Water underground
Insect screens
Massive biomass
Wind deflection
Trellis windbreaks
Insulation mounds
W-protection

Hot Wet Climate houses
High ceilings
Verandahs
Ventilated grids
Non sun-facing active rooms
Cross ventilation
Horizontal louvres
Roof cap
Fans
Trellis/shade houses
Orient for shade

Retrofitting is taking a house which is not environmentally designed and gradually turning it into a good environmental design. So for example, you can change how you live in the rooms, open up some areas for cool breezes, or close down some for warmth. You can add a solar hot water system or use biotecture to modify the climate. Your aims are to use less water, less non-renewable fuels and have a healthier, more comfortable house.

Student activities

- *Design a complete retrofit of the house you live in*
- *Design the least-consuming and most energy efficient house of your dreams*
- *Specify the climate*

UNIT 16
Zone I: Home food gardens

We can grow 80% of our food within 50 metres of our homes but we currently rely on conventional agriculture to supply our food needs. Agriculture as practised is the greatest threat to the planet because it causes excessive environmental degradation. Formerly land was a resource to live on and grow food and has only become a commodity since the commercialisation of food.

Ethiopia was a green paradise and many of our grains originated from there but the grains were stolen and cropping was directed at exports. The land was destroyed.

Have students learn to think in terms of three agricultures:

Agribusiness which is only concerned with food as a commodity in the international market, is export oriented agriculture. It relies on monocultures and high chemical inputs. There are also high capital inputs. Of its income, 30-40% goes to multinationals.

The greatest expense of agribusiness is servicing debt – not to pay off principal debt, nor improve land but to pay interest. The only way it is viable is with government subsidies.

Most agriculture falls under this heading.

Lawns require enormous chemical, water and energy inputs yet provide no food. Chemical fertilisers, fungicides and weedicides per hectare are double and sometimes treble those used in intensive agriculture.

In California, 14% of drinking water is used on lawns. In WA, 20% of clean drinking water is used on lawns. This is equal to 100" rain pa.

Lawns cost approximately $6.00 per sq.ft pa. to maintain, whereas a vegetable garden yields 15.00 UK pounds per square metre. A bed of annuals may cost around $300-$400 per square metre per annum to maintain. N.B. 1995 was the world anti-golf course year.

Home and community gardens are the only form of agriculture which directly feeds people. We can grow 80% of our food within 50m of our homes. In ex-communist countries, 60% of food for people was grown in home gardens. This constitutes 2% of agricultural land. For example, in former Yugoslavia in 1983, people were allowed 2-4 ha (5-10 ac.) per family. They were allowed to leave work at 3pm to work in their gardens. The State helped with deep-freezers and with mini-tractors.*

The yield from home gardens in the USA was estimated at $50 million pa in 1988. There is much additional information coming out daily in magazines, newspapers and the web. I add to my knowledge from these and collect planting calendars for most climates.

This is the unit many students have been waiting for, especially those who arrived thinking permaculture was only about growing chemical-free food. You may need to slant your lesson towards one of the following: tropical, cool temperate or arid climates, or urban populations or schools and community gardens.

By now your students are discussing ideas with each other and you, the teacher, and working in groups and in pairs and you are now working with a variety of teaching methods and continuing to use class expertise and interactive learning techniques. A practical period when students make a Zone I garden usually accompanies this unit.

Learning objectives

By the end of this unit students will be able to:

- Design a garden for food, herbs and flowers
- Make a sheet mulch garden
- Understand the role of herbs
- Select appropriate animals as maintainers and tractors
- Choose plants for permanence in food gardens

Teaching tools

- Designs or photos of permaculture gardens for different climates
- A range of design methods as they change according to size of garden
- Sides or photos to illustrate points in this unit
- Best of all: A mature permaculture garden to learn all this in

Ask a student to offer their garden for design and visit it first. Take criteria for a Zone I design, write them up and ask students to make a design working to these criteria. Then build all or part of the garden.

Make a Garden Fair with a number of activities e.g. A tyre pond, a herb spiral, a mulched garden etc. and groups of students take turns at different activities.

* Source: Morrow, Rosemary, visit to Yugoslavia in 1978.

With this unit students begin to redesign their home garden and by the end of Unit 21 will have a complete site and house design and totally retro-fitted their old garden (at least on paper).

Terms

Chinampa A garden system where fingers are dug out into a lake or canal to extend the edge to be close to water and modify the climate. Originally named in South America, now found in all delta systems.

Companion A plant that grows with others to protect or enhance them.

Edge The junction/zone that lies between two ecosystems or landscape and forms a border where materials or resources accumulate from each.

Keyhole Beds Beds shaped to allow gardeners to reach all plants without treading on the soil.

Sheet mulch Completely mulch a garden much like laying out a sheet and building soil.

Stacking Arrangement of plants to take advantage of all possible space and time, using tall and medium sized trees with a lower shrub and herb layer. It is a successional strategy. Care is taken so water and light competition are minimised.

Ethics

Make an abundant food garden to relieve the strain on marginal agricultural land and limit the risks of personal and environmental toxicity.

Improve productivityof the land you own or rent and leave something for the future.

Principles

Revise the concept of zoning and the objectives of Zone I.

A student nominates their home for the class to build a sheet mulch garden. You should visit it first. Don't do public land because it is rarely cared for.

Nutritional requirements of humans are for protein, energy and other nutrients. In Zone I, protein can be met through eggs, beans, cheeses, chickens, mushrooms and fish. Energy needs are met through staples, beans, oils and sugars.

Students think which garden vegetables can meet these needs.

Hot, wet climate drainage and water harvesting

Characteristics of Zone I

- Close to the house and is the area of most human activity where food eaten daily is grown
- Not more than 50m from the back door
- Intensively cultivated
- Vegetables, flowers and herbs as permanent as possible
- Often visited
- Use small animals as tractors, fertilisers and maintainers
- Sheet mulched
- Starts at the backdoor

Discuss the importance of diversity and permanence and the concepts of stacking, edge effect, keyholes.

Permanence in vegetable gardening has four components:

- Long term perennials
- Biennials
- Self-seeding plants
- Mulch

Needs, yields and inputs design method

Needs – list everything you want to do or have in the area. Each one is an element in the garden and each has needs and inputs.

- Herbs
- Flowers
- Recreation – exercise/play
- Work – e.g. Bicycle repair
- Clothesline
- Pond

Zone 1

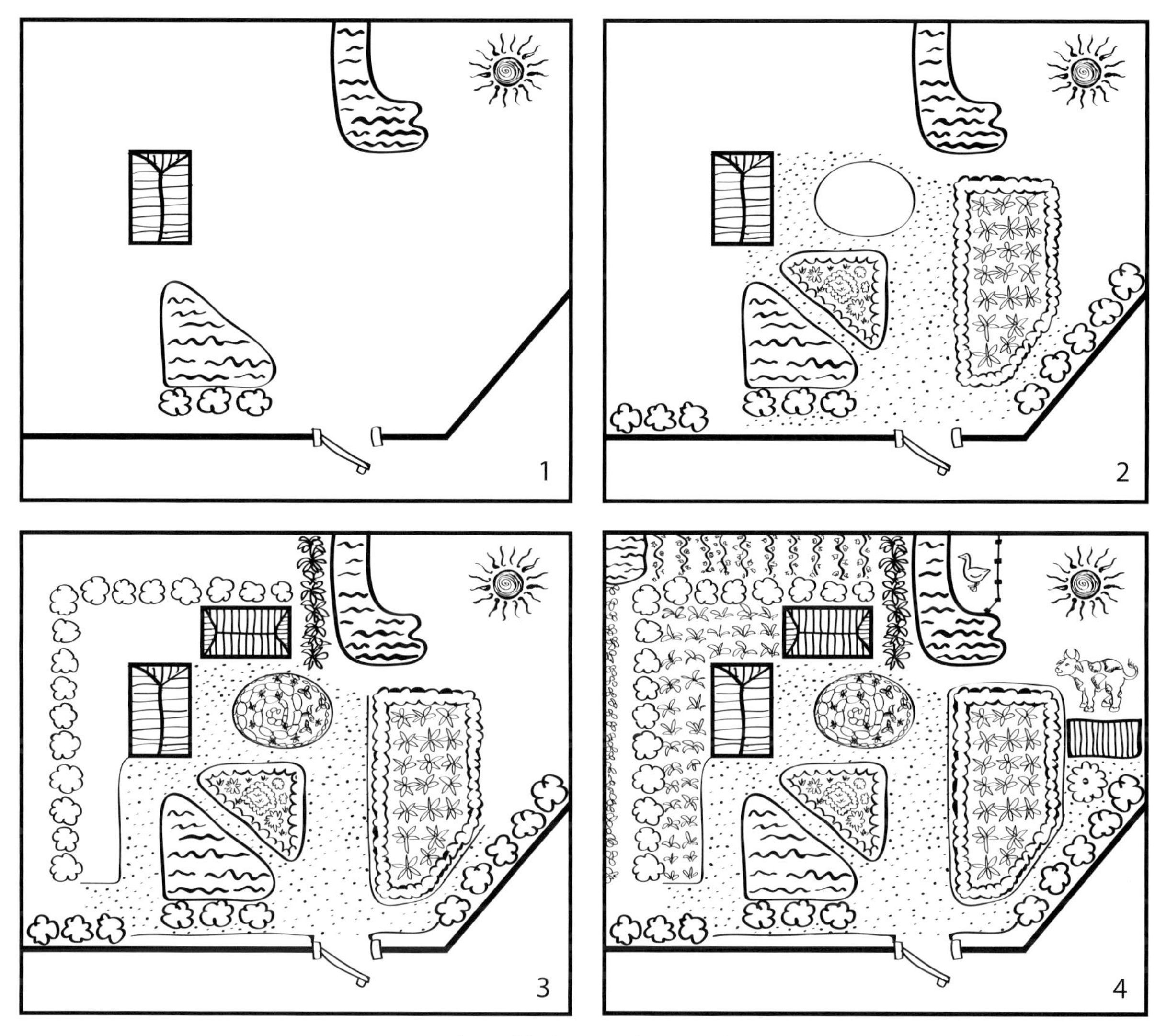

Putting a kitchen garden together

- Fruit trees
- Kitchen – lorena stove, BBQ Lorena stove
- Chickens on deep litter/tractors Streams
- Potting shed
- Outside loo (under vines to pee)
- There can be conflicting interests.
- Vegetables
- Compost bins
- Greenhouse
- Pets – dogs/rabbits
- Shadehouse
- Tool shed
- Pergola/trellis
- Water tanks
- Garden shed
- Wormbed
- Solar oven
- Bamboo – clumping only

Inputs – try to find second hand or recycled materials

- Water
- Nutrients
- Six hours of sun per day
- Protection from pests
- Seeds, cuttings
- Planting plan
- Mulch
- Old bricks, stones

Have the class design the whole garden area on a portable white/chalk board or poster paper based on inputs and needs. Make sure NEWW are supplied first – nutrients, energy, water and work.

Planting plan

Decisions about beds location and species selection are influenced by:

- Climate and soil
- Frequency of use and harvest
- Level of maintenance
- Lifetime of plants
- Growth habit
- Space required at maturity

When designing nutrition gardens you need to know what plants contain special minerals and vitamins. Draw up a chart on the board with the three headings: Area, Purpose/Use and What to grow. Using the following headings to fill out the table asking students for suggestions for the third column.

Stages of implementation

Eliminate lawns, especially small ones. Start with permanent circular paths. Start small even 1-2sq.m such as kitchen door garden or square metre garden and sack gardens. Group activities e.g. Tools and compost. Sheet mulch the whole area.

Sheet Mulching

The class builds the sheet mulch garden. Scavenge as much free material as possible to create a sheet mulch garden using:

- Compost
- Newspapers
- Hay/straw
- Sawdust
- Coffee grains
- Old timber for edgings etc.
- Manures
- Underfelt – not plastic
- Pine mulch
- Stable sweepings
- Bricks/stone

Discuss different weeds and how deep paper/ cardboard is needed and revise different mulches.

---

1. Slash grass, mow lawn and weeds and leave on ground.
2. First, wet the whole area and then paper or carpet the whole area.
3. Mark out paths with bricks or lime and cover the paths with sawdust.
4. Put scraps, manure, dolomite, compost on all other areas.
5. Then a layer of hay.
6. Finally place straw or other as top mulch.

Planting begins immediately

Students bring plants as cuttings, seeds and seedlings for the garden.

Start at the back door

- A lime or lemon tree with herbs such as chives and parsley underneath provides easy access for frequent use. Citrus store best on the tree.

Herb Spiral

- Use culinary herbs as close as possible to the back door for easy harvesting. Interplant annuals and perennials – place according to microclimate i.e. thymes facing the hot west, over rocks. Herbs to be used can include:
- Marjoram and Sage – garden and Mexican
- Rosemary – two forms
- Basil – Bush, Sweet, Opal, Cinnamon for marinades; Lemon for fish poultry and deserts; Mammoth for stuffing; Well-sweep Miniature Purple, Napoletano for salads; Camphor for bees; Genovese, Piccolo Fino Verde, Lettuce Leaf for Italian cooking; Thai, Thrysiflora, Anise for Asian; Spicy Globe, Bush, French Fine Leaf. Basil needs warm day and night temperatures. Start indoors about three to four weeks before last frost
- Savory – Winter, Summer
- Oregano – Golden, Greek, Garden
- Thymes – Wooly, Turkey, etc
- Tarragon – French, Russian

Clipping Beds

These edge the paths and the inside of the keyholes. They are as permanent as possible. Plants are constantly harvested, some on a daily basis. All have extremely high mineral content of calcium and vitamin C. As they are constantly harvested, keep adding worm castings, potash and lime. Useful plants include:

- Chives – garden, garlic
- Cress – garden, upland
- Dandelion
- Salad burnett
- Nasturtiums
- Parsley – curly, Italian, Japanese
- Sorrel – French
- Corn salad
- Mustard greens

Plucking Beds

These are pathside and are fast growers and can quickly become too big or those which are frequently harvested without removing the whole plant.

- Broccoli
- Dill

- Swiss Chard
- Capsicum
- Coriander
- Bunching onions
- Non-hearting lettuces
- Silverbeet
- Zucchini
- Kale Leeks
- Brussel sprouts
- Celery

Use several varieties of each species

Narrow Beds

These are narrow to maximize the plants' solar access. Plants include:

- Carrots
- Radishes
- Beans
- Okra
- Tomatoes
- Peas
- Asparagus
- Eggplant
- Planting radish and carrots, mix 60% sand with 30% radish and 10% carrot seeds. Make 1/2" depression and sprinkle the mixture. Radish acts as a pioneer, keeping weeds down, ripen first and ease ground for carrots. Radishes are fast growers, carrots are slow. Basil and parsley grow well with these.
- Planting asparagus is a 20 year investment, grows well with tomatoes, plant in trenches and gradually fill in. In some areas, as asparagus is harvested, plant tomato seedlings. Harvest asparagus for 6-8 weeks then let it grow to renew its rootstock for next year.
- Globe Artichokes from seed, or seedlings. Good for seven or more years. Perennial. Thin out at about three years. Require good feeding, susceptible to slugs so use fresh wood-ash to protect the young plants.

Broad Beds

The plants here are slower growing, one-off, usually annual crops. They do not require as much attention and include:

- Hearted lettuces
- Cabbages
- Winter lupins
- Sweet corn plant no later than early December. 16 plants 4 x 4 grid planting is good for pollination. Varieties are Platinum Lady, Tuxedo, Butter and Sugar (two colour),
- Pumpkin, select for vigour, stress resistance and flavour
- Globe artichoke
- Chinese cabbage
- Swede
- Turnips

Broadscale grains as staples

- Jerusalem artichoke
- Cauliflowers
- Beetroot

Many gardens do not have enough space but where there is, include broadscale grain crops grown Fukuoka style and include:

- Corn
- Wheat
- Rice
- Potatoes – type of mulch is critical to yield, unmulched potatoes yield 7kg with leaves, straw, grass mulches yield is 20-22.5kg
- Corn, maize, gives the highest yields and 0.5ac of corn will give staple self-sufficiency

Fence and trellis crops

Fence and trellis crops increase the growing area, especially in small gardens. All existing structures and new ones such as tipis can be used for vertical growing. Trees such as citrus can also be used as structures. Outdoor structures allow great creativity, for instance, climbing plants can be grown over ponds. Some trellis crops are:

- Passionfruit – banana, black, yellow
- Beans
- Grapes
- Cucumbers
- Boysenberries
- Peas
- Jicama
- Pumpkins
- Loganberries
- Chokoes
- Kiwifruit
- NZ spinach

Additional points

Crop rotation

This is moving different major crops around the garden to follow each other. It can save space, cut down on pest problems and balance soil nutrients. There are traditional rotational planting cycles and one of these is: Legumes –––> Corn etc. –––> Leafy vegetables (brassicas) –––> Root vegetables –––> Legumes

Nutrients

The main nutrients come from compost made from kitchen organic waste, from worm farms and from small caged animals such as rabbits, a chicken or two in a tractor, guinea pigs, and quail which are rotated around the garden as needed.

Water

Household grey water from showers and washing machines can be used. Water is passed through a duck weed pond for cleaning. Use only pure soap in the house, except in alkaline desert areas where it is possible to use high phosphate laundry washes. Grey water is safe if it is used under 5.0cm of mulch such as straw and is in contact with soil.

Weed reduction

These are best managed with fully sheet mulched beds and close planting so the leaves of the plants are just in contact with each other. Use paths of ash, sawdust, brick (especially good in cold climate) and stone with yarrow, chamomile, thyme, or clover. Weed barriers with plants such as lemon grass planted along neighbouring fences will inhibit the invasion of weeds.

Companion planting

This works when some plants enhance or protect others. There are many combinations. Here are some:

Asparagus + beans + basil + tomatoes
Radish + carrot
Corn + beans or lupins

Don't plant onions and beans together.

Hand out a table of companion planting.

Useful non-food species

Halibash gourds – for food containers, cups.
Loofahs – to replace washers, sponges.
New Zealand flax – for strong garden ties.

Herbs

Plant herbs throughout the garden for pest control, medicinal, cosmetic and culinary uses. Garlic is one of the most potent. The more pungent the herb, the more effective it is as a control. Break pieces of leaves of such herbs as tansy and yarrow while walking through the garden. Keep mint in pots, it spreads too greatly. Grow herbs for medicinal and cosmetic manufacture at home. Also sell to local markets, and/or greengrocers.

Native plants in Zone I

Grow native shrubs, herbs and grasses because they draw predators of many pests and some provide foods while adding to biodiversity and keep the gene pool alive.

- Lilli pilli
- Quandong
- Grevillea (Honey gem)
- Small bottlebrush
- Appleberries (*Billardiera scandens, B. longiflora* and *B. cymosa*)

The question of fruit

Fruit in a small home garden should be carefully selected for appropriateness and regular supply over the year. Draw up a calendar of yields.

- Citrus – Eureka lemon – good standby, long yielding
- Stone fruit – nectarines, apricots, peaches
- Small fruit – berries and brambles and currants
- Rhubarb – strawberries
- Vines – grapes, kiwifruit
- Espalier – use fences, sheds and walls. Use dwarf stock
- Multigraft trees – e.g. Jonathans or Granny Smith apples

Mr Dawson Burns, an orchardist of Tasmania, stripped a large red delicious tree in 1980 and started grafting other varieties onto it. By 1989 he had 230 new varieties on it. Grafting wood has been sent from all over the world, including one from USSR. His commercial orchard is a tourist attraction for coach tours. The first apples mature in January and the last one is eaten in August.

Oddments

Don't have pine trees near the vegetable garden. They rob soil and shade out plants.

Keep some beds for acid-loving plants such as rhubarb, use lime or dolomite on the others. Use a calendar to plan continuity of food supply. Concentrate on planting times, not on harvesting or you won't have your plants in on time.

Keep rabbits, quail, guinea pigs in cages to recycle waste, provide manure and eat down grass instead of mowing it.

Keep poultry out of the garden unless they are caged because their feet are too destructive.

One Khaki Campbell duck can be OK, but do protect young seedlings because they like green shoots.

Student activities

- *Redesign (retrofit) your garden, or garden-to-be according to the principles discussed in this unit*
- *Draw up a harvest calendar of how your garden will supply you with vegetables and salads all year*
- *Draw up a calendar of regular fruit supply*

UNIT 17
Zone II: Orchards and food forests

As people become more aware of biocides in their environment they want the security of clean fruit by growing their own. Clean fruit will increasingly take up more of the present market. They also want better flavour and a wider range of species.

Huge amounts of fruit and nuts are imported. Many use unfair labour, pesticides, and other chemicals, even irradiated before entering countries. Most commercial fruit is gambled by the futures markets. Commonly, monocultures have replaced small scale farmers who grew food.

Permaculture orchards are areas of intensive production and conceived of as food forests. That is, they contain a diversity of food and non-food trees, mixed to encourage predators, and confuse pests. Our model is a forest with every plant having productive uses. Every garden can have some fruit.

In the original permaculture syllabus the emphasis was on cool temperate orchards with their need for space and light. Since then there has been an expansion of knowledge and experience in designing food forests for all climates, including cool temperate. Monsoonal and hot wet climates have always naturally grown food forests and theirs is our model. For example, a natural guild is coconut trees, pepper plants and citrus beside fishponds.

We can think of Zone II as a fruit forest, and Zone III can be a food forest if it provides commercial quantities.

In this unit students revise the ideas of guilds and forests where they learned about the needs of trees for:

- Pollination
- Pruning
- Fertilising
- Seed dispersal
- Pest management

In a natural forest this occurs through interactions and balance. Through design we want many of these functions carried out by other plants or animals.

Students have a few minutes to discuss how to make a fruit orchard function as a forest. The model here is mainly for warm temperate climates.

Learning objectives

By the end of this unit students will have:

- The theoretical basis to design and implement a design for an orchard
- Sufficient knowledge to carry out further research

They will be able to:

- Select and prepare a site for an orchard
- Retrofit an old orchard
- Supply the needs of an orchard for: water, nutrients, protection, energy, and work requirements
- Select appropriate species and ensure continuity of supply

Students have already redesigned their land to implement their Zone I garden. In this unit they will design a Zone II fruit forest. They will need their plans of their land.

Teaching tools

- Revise sector analysis and windbreak design and discuss the need to modify these in different bioregions and specific sites
- Plan of orchard illustrating soil and water from a water harvesting scheme which works by gravity then place the fruit trees on the plan
- Ask students to design a fruit forest for themselves as you go through the lesson and cover site selection, preparation, water, species selection etc. Gradually build up their own plan
- Have a fruit calendar ready for discussion
- Revise the waru or guild as a basic concept covered in the unit on forests

Terms

Food forest A forest of nearly 100% productive species.

Grafted species Plants with a strong root systems onto which is grafted a cutting from another plant which is high yielding or disease resistant.

Orchard A food forest with many fixed and mobile elements, e.g. Trees, pollinators, pruners, fertilisers.

Ethics

Plant productive fruit and nut trees for now, and for the future to return what has been removed. Every plant in a food forest is productive.

Principles

- An orchard is a waru or guild of interrelated and interdependent fixed and mobile elements which work for the trees and the trees work for them
- The diversity of trees and animals is the basis of this cultivated ecosystem
- Productive trees form the basis of a sustainable system and provide excellent return for effort
- Growth is more vigorous and pests and diseases are greatly reduced when a 'forest' of food is created
- The ecosystem is in balance
- Losses and failures are reduced by thoughtful observation and knowledge of local conditions

Begin with the principles of succession

- Start by planting nitrogen fixing species to enrich the soil
- Interplant throughout this first flush forest with the climax species when the nitrogen fixing species are tall enough to give protection to the new fruit trees
- Harvest the nitrogen fixing species for firewood, mulch or fruit
- Harvest fruit each year and graft them progressively
- Replant every year and increase planting to rebuild the fruit forest

Layered food forest

Characteristics of Zone II orchard

- Intensively cultivated
- Heavily mulched
- Well maintained
- Selected and grafted species
- Densely planted
- Multi-storied and stacked
- Multipurpose walks to collect eggs, mulch, fruit etc

Needs of a Zone II orchard

An orchard is healthy and yields heavily when its needs are met. So the design method we use here is for Needs, Yields and Functions/outputs? It needs:

- Water
- Energy
- Pollination
- Harvesting
- Access
- Soil nutrients
- Wind protection
- Pruning
- Pest management

Meeting its needs

To meet the needs of the whole orchard, first meet the needs of the trees and plant these. Then later introduce the animals.

Have your students site and draw these on their land.

Protection

Orchards need fire, weed and wind protection. After doing a sector analysis, design and plant appropriate windbreak species including productive ones and those which entice predator birds. Coppice windbreak plants to thicken them and provide mulch. Concentrate on local species that thrive and add more difficult ones later when micro-climates are established.

Barrier plants against weeds are mainly herbs, e.g. Lemongrass, comfrey which compete with weeds around trees.

Ringlock fences are better than barbed wire.

Access

Use high contours for tractor/vehicle access roads. These give good visibility of the orchard for monitoring.

Water

- Use high dams and water for gravity-fed strategic watering

- Increase soil water through ripping for three years along the contours
- Nesting sites, hollow logs
- Build small ponds for frogs, ducks, geese for pest control

Drainage is important, and even on flat lands there are natural drainage points. Create swales at these and use ripping to drain water into swales.

Very few trees can take extensive waterlogging. One week is about the maximum. Quinces, pecans, mulberries and citrus are most resistant to waterlogging – in that order. They are good for flash-flood lowlands.

In very wet areas, you can create diversity by building chinampa-style systems. e.g. Almonds, apricots, figs need good drainage.

Nutrients

Remember to check for indicators for the health of plants. For example, the invasion of bracken can mean low nitrogen. Nitrogen-fixing trees such as acacias, broom, alder and leucaena are very important because they can supply up to 370lb of nitrogen per acre. Other crop legumes, such as clover and lupins also supply nitrogen. Herbs mine lower soil horizons for minerals.

Comfrey in summer can be cut 3-4 times to ground level and used for liquid fertiliser or mulch.

Poultry and pig manure from these foragers complement the nitrogen-fixing and herbal ground cover as nutrients. Ants compost effectively by taking organic matter below ground, where plant roots can extract nutrients.

Rain is like a nutrient soup running off trees.

Mulching

Ground cover mulches are living mulches in summer and dead mulches in winter. Plants return 25% of themselves to the soil each year and all contribute to soil nutrition. Living mulches are grown as ground covers. Seeds are broadcast through the system, for example, through the dead summer crop which is later slashed.

These mulches will add the vital organic matter and hold water in the soil where the plants can use it. They will also help soil micro-organisms. The desired goal is biomass not rampancy. Orchards also need ground covers to control competitive grasses. Mulches also help to control codling moth.

Mulch plants include cowpeas, soybeans, legumes. Under fruit trees start planting onions, chives, comfrey, clovers (red and white), lupins, pumpkins, vines and include chicken forage plants.

Pollination

Pollinators are any agents which transfer pollen. They can be wind, water, insects, wasps, birds and bees. Most insect pollinators are encouraged by flowering plants. Some, like bees, do not function when it is very windy, so there is a great need for orchard windbreaks.

Bees are primary agents for cross-pollination of such fruits as apples, almonds and hazels. The prefix Mel is Greek and Latin for honey and often a guide to plants that attract bees e.g. Melalot, a white and yellow flowering plant good for bees in orchards. Borage is excellent and so is lemon balm.

Orchards are floristic systems, where every plant should flower. Flowers can be stacked into orchards really well. Plant a range of herbs and flowers such as fennel, borage, dill and carrots are excellent. Nasturtiums, daffodils, jonquils, hyacinths, irises all grow well in association with fruit, and flower early for bees. Melaleuca, teatree, bottlebrush, Pittosporum, Photinia, borage and lemon balm are all useful. Sow local wildflower seed among the trees.

Pest management

In a monoculture a crop can be 100% pest affected. While there will be some damage in polycultures, if they are well-designed, damage can be limited to 4%. This gives the benefits of chemical-free food and no financial debt to chemical companies.

Orchards must be well maintained to decrease susceptibility to disease and pests.

Fruit trees, like all others, need foragers. Pigs and poultry such as guinea fowl, hens, geese, and ducks clean up rotten and diseased fruit which harbour orchard pests, fertilise trees, and mow and weed the orchard.

Six geese maintain 0.5 ha (l acre) of orchard nicely.

Encourage frogs, loose rocks for lizards around ponds, ducks, predator birds and insects. Fukuoka grew Tasmanian blackwoods (*A. melanoxylon*) in his orchard as decoys to protect against aphids which prefer Acacias. Birds are natural predators of aphids and they breed up and control the aphids.

Fruit and berries growing in suntraps and windbreaks encourage robber birds, which usually prefer watery fruits before quite ripe, and attack sour fruit when green. Plant ungrafted, less-sweet species in windbreaks to divert birds from fruits. Hawthorn, lilly pilly, Persoonia, elderberries, capostrum, Irish strawberry tree (good preserves), Kaffir plum, Cotoneasters, crab apples, medlars and kangaroo apples are good choices, as are many indigenous bush foods.

To deter birds, confusion techniques such as kites,

fishing wire and aluminium, will only work if put into action before birds discover the fruit. Take them down quickly once fruit is harvested. Nets work well in home gardens and it is good to have a small orchard all under netting. Dwarf trees are easier to net. Hanging small rubber snakes in fruit trees and moving these every few days is said to be effective. Don't over-fertilise as too much fast lush growth attracts insect pests.

Pruning

Pruning branches leaves trees susceptible to disease and insect attack, so it must be done carefully and selectively. Some apples and pears, if pruned annually, bear biennially. However if pruned only for disease, they bear more evenly.

Persimmons, loquats, feijoas and figs do not respond to pruning and may be damaged, and apricots and almonds don't much like being pruned.

Some trees bear only on new wood, others on last year's growth. Mulberries produce on new wood so cut them back after summer bearing. They coppice well and the prunings are good fodder for animals.

In general, deciduous trees, if pruned at all, are pruned in winter; others after flowering. Fertilise trees after pruning. Pruning is most appropriate in Zone I to limit size and height of the tree and in cities or where espaliers are desirable.

In Zone 1, pruned fruit trees, such as apples and pears, can make very attractive living fences if branching out shoots are nipped out while they are still buds. The yield can be nearly as good as unpruned orchard trees because of available sunlight, and the energy of the plant goes to fruit rather than twigs and branches.

In the first year, harvest the first fruit early so the tree is not stressed.

Species selection

Use grafted stock and start with a small orchard of many different varieties, selecting what does well in the local area. Start with one appropriate variety and add other varieties which bear later in the year. Work out, or locate, harvest calender for each bioregion. Select a few peripheral species to take advantage of either climate change or non-average seasons. Concentrate on local varieties and their needs for:

- Light requirements – frost needs
- Time of leaf drop – blossom set
- Adult size

When selecting species, consider not only the climate of the region but the more specialised micro-climates, such as different sides of a house or hill. Some fruit tree peculiarities are:

- Loquats will flower and bear earlier if the micro-climate is warmth enhanced i.e. can steal the early market
- Chestnuts need a 1-1.1/2m depth of sandy loam or loam and suffer if there is a high water table
- Persimmons cannot stand wind. Citrus will take some wind but then bear mostly on the lee side
- Mangoes and some seedless mandarin can grow to 10-15m high and function as windbreaks as well
- Avocados like well-drained soil, high humidity and thick mulch
- Kiwi fruit prefer an easterly aspect
- Grapes and wisteria are good on northerly aspects since they drop their leaves early in winter. Kiwi fruit are good in hotter drier areas, where there is sufficient soil-water. Dwarf species mature about 2m high and grow well in tubs

Many fruit trees are cross-pollinated and some will cross with others, for example, almonds/peaches, and almonds/apricots.

Apples can turn into quinces, and pears to apples, most apple trees are cross-pollinated, and so are almonds. Some plums and pears are cross-pollinators.

Look up orchard species for special climates. Plant native plants where possible e.g. In Australia, these are quandong, fig and cumbungi.

Cool temperate species

Apricots – trevatt
Plums – narrabeen, greengage, bloodplums
Bells is an early November peach.
Apples – gravenstein, Jonathan, Granny Smith, delicious, northern spy (very old), King David, Rome beauty, cox's orange pippin
Some plants require a chill period, e.g. Apples require 110 hours of frost, so place these carefully.

Dry summer species

Olives, figs, grapes, herbs, carobs – all good in cool climates as well but are frost sensitive in the first few years so plant wattles with them, or even 1-2 years before. After this they are hardy and very drought resistant. In summer grow melons and pumpkins. (Chinese gardeners grow winter vegetables and leave fallow in summer.)

Tropical orchard species

Monstera, mangosteen, Butia palm, herbs, coffee, tea, tamarillos, mangos, cashew, avocado, bananas, papaya, guava, chicku, citrus, coconut palms and pepper, and many more.

Rolling permaculture

Improving an existing orchard and changing it to a permaculture design is called 'Rolling Permaculture'.

First – rip with agro-plough then build up ground covers of clovers, lupins, comfrey etc. to add organic matter to the soil. Broadcast buckwheat, turnips, radish, daikon, root crops including potatoes.

Secondly – establish windbreaks along the following design. Close the windbreak on the prevailing windside, and check windbreak species do not escape and become rampant. They will function to filter dust, bee fodder, fuel, bird habitat, mulch and additional animal fodder.

Windbreak Design for 5-row Orchard

Most important is to get fast-growing acacias in the first row within one year to afford protection to the rest. In very acid soil add lime or dolomite.

In-crop planting of nitrogen-fixing trees at 3-5 per quarter acre, use acacia, tagasaste, carob, false acacia etc. as appropriate.

Meanwhile continue to establish best conditions for soil, wind protection and planting. If soil is compacted, rip and plant to melons, pumpkins, potatoes, clovers and comfrey.

When planting, use thick newspaper or underfelt mulch. Kikuyu requires very thick mulch but can be shaded out. Globe artichokes are good for keeping weeds down. Spot mulch an established orchard and work out from the trees, plant herbs close to the tree trunks.

Caution

Bracken, pines, walnuts (a rare monoculture crop), eucalyptus should not be planted in orchards.

Student activities

- *Design an orchard for where you live – it may be suburban or on acres*
- *Do construction drawings, by putting in the actual species and their spacing. Remember that about 6m intervals is appropriate*
- *Develop a harvest calender to show how it meets your need for fruit all year*
- *Plan to harvest a crop from the nitrogen fixing plants and interplants used while trees are maturing*
- *In a food forest, chop and drop competitive species to enable light, as the most limiting factor, to be adequate for fruit production*

What is Edible Forest Gardening?

Edible forest gardening is the art and science of putting plants together in woodland like patterns that forge mutually beneficial relationships, creating a garden ecosystem that is more than the sum of its parts. You can grow fruits, nuts, vegetables, herbs, mushrooms, other useful plants, and animals in a way that mimics natural ecosystems. You can create a beautiful, diverse, high-yield garden. If designed with care and deep understanding of ecosystem function, you can also design a garden that is largely self-maintaining. In many of the world's temperate-climate regions, your garden would soon start reverting to forest if you were to stop managing it. We humans work hard to hold back succession—mowing, weeding, plowing, and spraying. Why not put up a sail and glide along with the land's natural tendency to grow trees? By mimicking the structure and function of forest ecosystems we can gain a number of benefits.

Why Grow an Edible Forest Garden?

While each forest gardener will have unique design goals, forest gardening in general has three primary practical intentions:

- High yields of diverse products such as food, fuel, fiber, fodder, fertilizer, 'pharmaceuticals' and fun
- A largely self-maintaining garden and;
- A healthy ecosystem

These three goals are mutually reinforcing. For example, diverse crops make it easier to design a healthy, self-maintaining ecosystem, and a healthy garden ecosystem should have reduced maintenance requirements. However, forest gardening also has higher aims.

As Masanobu Fukuoka once said, 'The ultimate goal of farming is not the growing of crops, but the cultivation and perfection of human beings.' How we garden reflects our worldview. The ultimate goal of forest gardening is not only the growing of crops, but the cultivation and perfection of new ways of seeing, of thinking, and of acting in the world. Forest gardening gives us a visceral experience of ecology in action, teaching us how the planet works and changing our self-perceptions. Forest gardening helps us take our rightful place as part of nature, doing nature's work, rather than as separate entities intervening and dominating the natural world.

UNIT 18
Zone II: Food forests and small animals

This unit is not a comprehensive study of the husbandry of small animals. It continues the principles and concepts used in Zone II plant systems and integrates the two components of Zone II; plants and animals. It is the next area of intensive cultivation after Zone I and requires more water, nutrient, human input and energy sources to implement. On the other hand, in time, it needs fewer inputs with increasing yields.

Forest gardening is a low-maintenance, organic plant-based food production and agroforestry system based on woodland ecosystems, incorporating fruit and nut trees, shrubs, herbs, vines and perennial vegetables which have yields directly useful to humans. Making use of companion planting, these can be intermixed to grow in a succession of layers, to replicate a woodland habitat. en.wikipedia.org/wiki/Forest_gardening

Orchards and small animal systems meet each other's needs. The animals' needs for free ranging foods, water, and shelter. Chicken housing should be within easy access of an orchard and/or a tractor- vegetable garden. Chicken and/or pig tractor systems are the only really effective ways to grow organic vegetables and fruits commercially which if they were sufficiently extensive and commercial, would then be called Zone III – the zone of income earning agriculture.

Suitable animal species for food forests are: chickens, ducks, geese, turkeys, pigs and you must have bees for fruit tree pollination. There can also be specific local indigenous animals.

This unit introduces principles of growing chickens, geese, pigs and bees in orchards.

Learning objectives

By the end of this unit students will be able to:

- List the needs of poultry and how they are satisfied
- Design healthy and humane poultry housing
- Use animals to enhance or support other enterprises
- Describe the basics of beekeeping
- Describe the husbandry of other small livestock

Teaching tools

- Photos and designs for animal housing for different species e.g. Chicken, duck shed and bee hives
- Designs showing where to place the animal housing so nutrients and water recycle from them, with least work and maximum effectiveness, to the fruit trees
- Students by now have made a plan of their own home and made it more energy efficient
- They have designed a new 'permaculture' home garden, and sited an orchard with its permanent plant species
- In this unit they add the animal components

Revise the design exercise using the need and yields analysis method.

Terms

Grains Also called staples for animals, are wheat, oats, barley, corn, rye, millet, rice.

Grit Coarse materials such as sand, eggshell, required by all birds in their crop to help with digestion.

Maintain Using animals to keep weeds, diseases and pests at an acceptable level in any crop – not removing all and keeping soil from being exposed.

Stocking rate or **Carrying capacity** The number of animals kept on a certain sized piece of land e.g. 9/ha. while maintaining it in a good state.

Tractor Using animals such as chickens or pigs to completely clean up an area of land – turning it over, fertilising it and removing weeds.

Students consider one animal they already have or would like to raise and then answer these questions about introducing them into a plant system.

- *What does it eat? How does it function? What are its products?*
- *Is the species suitable for the climate or is there a locally adapted one?*
- *What impact will it have on the environment?*
- *What size area will it need?*
- *What are its husbandry needs? Who is prepared to take this responsibility?*
- *What are the owner's needs and tastes?*
- *Is there a reasonable market demand?*
- *What are their breeding habits? If you don't want young, don't buy pregnant.*
- *How do they interact with other animals?*
- *What diseases are they susceptible to?*

- *How will they integrate with other farm functions?*
- *What are their other uses?*

Ethics

Create an orchard as an integrated, sustainable ecosystem of productive species.

Integrate animals harmoniously in a fruit forest. Natural forests are our models.

Principles

- Plants meet the needs of animals for food, shelter, moisture in fruit, and medicines
- Animals provide plants with nutrients, pest control, soil conditioning
- Other animals are encouraged to pollinate, fertilise, and prune plants
- Plants and animals grow best when adapted to local areas

Design method: Analysis

The design method used placing animals in growing systems is that of analysis which describes how the needs and yields of the animals are matched to the perennial plant species needs and yields, and these are integrated with all other aspects such as water and landform.

Some designs for managing chickens

- Square yards with housing in the centre and animals allowed into different sectors
- Animal house with an alley and off it are small grazing lots
- Linda Woodrow chicken domes*

Chickens

Use orchards to:

- Grow a small number of chickens maintaining high stocking rates and rotate them often – requires
- Mobile fencing
- Produce eggs or meat in commercial numbers using large scale set-ups
- Aim to grow 90% of the chickens' dietary needs

Needs for chickens in orchards: The 4 Gs for food

Grains Grow some, such as wheat, oats, barley, corn and amaranth. Many can be simply broadcast through an orchard after the chickens have tractored it. Also seed dropped from trees such as tagasaste, acacias, lucerne, oak, honey locust, carob and leucaena. Some seed such as carob and honey locust may need to be hammermilled.

Greens Such as comfrey, clovers, chicory, oxalis, chives, parsley, dandelions, cleavers can be broadcast. Plant fruit and berries such as: vines as grapes, chokos, black and banana passionfruit and kiwifruit suit chickens. Mulberries, hawthorn, elderberries, sunflowers, figs, guavas, loquats, tamarillo, cucumber, and pigeon pea, bananas and paw-paw go well with chickens. They will eat any fallen fruit including those that may have fungal or viral diseases.

Grit Sand and crushed shells (roasted) keep digestion functioning well. If animal is ailing, feel the crop and see if it is full. If not, why?

Grubs – insect protein If ranging freely, there should be enough insects and grubs for chickens not to require protein supplements. Bantams, being more carnivorous are very good at cleaning up fruit fly larvae. Chickens love termites. Attract termites with newly cut pine timber, or corrugated cardboard. And, unlimited fresh water at all times.

Housing, health and behaviour needs of chickens in orchards

Show plan view and elevation of different chicken housing.

Chickens are descended from forest foragers of SE Asia and Burma and need those environmental conditions. They are creatures of habit. They like to nest at home. Let them out after midday when they have laid their eggs. Make sure they always come when you call by giving them a small amount of grain each time.

Consider:

Climate: Face housing to the sun and door to the moderate east
Safety: From foxes, dogs and hawks
Health: Ticks/lice, diseases

Design for health

Medicinal plants Oxalis, clovers, wormwood, mugwort, dandelion are good for keeping chickens healthy. Mugwort and wormwood can be grown as a hedge and are reputed to help repel lice and ticks. Onionweed, nutgrass, couch and kikiyu are also cleaned up by chickens.

Dustbaths Very important as they keep them free of parasites.

Paint all construction timbers with sump oil and

* *The Permaculture Home Garden.* Linda Woodrow; Penguin Books, Australia, 1996

pyrethrum. Use derris dust and lime sawdust on the floor. Lime the straw yard every six months for fowl lice. Bamboo leaves are good as lice repellent. Happy and healthy chickens will not get sick and housing is part of this.

In cool climates, insulate the roof because severe cold can kill chickens. On very wet days keep them in a straw yard or they will not lay. It is very important for chickens to have a straw yard so they do not go directly from roost to range. The house must be dry and well-drained with adequate sun.

When building a chicken house, let the manure build up until there is about 15cm (6in) then compact it down so there is a 5cm (2in) layer of cement-like surface. There will be an antibiotic substance in this that help in preventing disease. This is also true for pigeons.

Design for safety Where foxes are common, net the whole yard. Also dig wire into the ground and turn back so dogs, foxes, goannas cannot get under it. Have floppy wire at the top so chickens can't perch and fly out and foxes are not comfortable entering.

To deter hawks and eagles, grow trees in the yard to disrupt their flight paths.

Design for companionship Place all roosts at the same height to stop competition and promote co-operation in the hen house.

Hen behaviour and numbers Chickens are social animals and need company. Their social orders are quite strict. Ideally there are a dozen hens to one rooster. Over 20 per rooster and flock behaviour breaks down. The optimum is 15 hens per rooster. Two roosters and 35 hens will happily co-exist. Where there are two roosters, place their roosts at equal heights at opposite ends of the henhouse. Do not have more than 35 hens and roosters to a pen. This means that 100 hens ideally require four pens.

50-100 hens will maintain a 0.5ha (1.1ac) orchard extremely well. They will keep weeds down but won't eliminate them. 500 per ha (2.2ac) will eliminate all growth except for trees within a few weeks. By rotating 100 commercially raised hens, so they graze about 1/20th of a hectare at a time, they can be made to function to the equivalent of 500 hens per ha tractoring (cleaning up and full manuring an area which has just finished growing organic vegetables). Move them on when they have cleaned up and manured the area. (They will also clean up cockroaches).

A few hens can be allowed briefly into an established vegetable garden for 1/2 hour before sunset, and quickly enticed out with grain.

Hens allowed to forage in a windbreak or firebreak will keep the ground clean to reduce fire hazard.

Products Surplus meat is eaten so humans or other animals act as natural predators. Hens can live for up to 20 years. Preserve eggs in big crocks of paraffin oil.
Breeding: White birds for hot areas; black for cold
Eggs: Black Australorps, Rhode Island Red
Meat: Cornish, Plymouth Rock
Both: Suffolk White. Light Sussex
Other: Chinese Silkies, Indian Game

Keep chickens which are docile, have a red comb and are long laying. Cull those which are aggressive, have a pale comb or are weak.

Ducks

Don't house ducks and chickens together. Ducks like wet, sloppy conditions. Chickens prefer dry.

Ducks are hardier than hens and can withstand colder, damper conditions and are less susceptible to disease. They are especially good in aquaculture systems and will be further discussed in that unit.

Some breeds e.g. Khaki Campbell and Welsh Harlequin will outlay hens. Three duck eggs are equivalent to four standard sized hen eggs. A duck's laying life is two to three times greater than a hen's. Ducks eat more than hens and are more efficient scavengers, especially if a pond is available. They lay well on a more varied diet. They are efficient snail managers.

Generally they are less destructive in gardens and will keep down pests. They will harm young seedlings.

They control slugs and snails most efficiently.

Breeds

- Black East Indian and the Cayuga ducks hatch and rear own young and have superb game flavour
- Muscovy, the 'mother supreme', will hatch anything
- Khaki Campbell is the top egg laying breed and a great slug eater
- Aylesbury is the top table bird
- Welsh Harlequin is a good dual purpose bird
- Indian Runner are good egg layers and good slug foragers
- Housing

Must be predator-proofed. Place on islands, or, mesh a water house attached to a land house.

Geese

Geese originated in swamps and wetlands. Most require water to breed satisfactorily. The main function of geese is to weed grasses and they are an excellent alternative to herbicides. They love narrow-leafed plants and are very good for weeding between broad-leafed crops.

6-12 geese per 0.5ha. (1.1ac) is reasonable for maintaining grassland.

There are Chinese geese, as well as European geese. The Chinese geese are less aggressive but not such good watchdogs. It is said that goslings can be trained to a coloured flag by keeping the flag over the feed bowl. The geese will then follow the gooseherder and flag to the grazing and stay there till it is time to follow the flag home or as long as the flag is pegged in the ground.

Pigeons

Originated on the rockfaces of the deserts of N.E. Africa.

Needs Grit, large seeds, maize, peas, beans, hemp.
Yields Pigeon manure makes a high quality liquid manure. It is also a powerful insecticide and 4mm sterilises animal pens such as henhouses.
Homing If trained, pigeons will fly 100 miles with messages.
Husbandry Keep locked up until the first brood is raised.
Housing There are many designs. Build up floor and collect manure. Use roof to collect water. Provide the flight section with landings and roosts. Keep free-range in large covered cages.
Predators Rats love pigeons so their housing must be rat-proofed.

Bees

The survival of bees is seriously threatened and the main reasons given are:

- Chemicals in the environment
- Monocultures
- Reduced food for bees
- Increase in pests and diseases of bees
- Reduced resistance to pests and diseases

Integrate bees into every farm and garden design. A whole farm could be planted to bee forage and still supply other functions. Logically, honey is the appropriate sugar for cool climates. Importing cane, palm or beet sugar is not a good permaculture strategy.

Ask why?

Sugar beet requires large areas and a cool climate and is grown mainly in the northern hemisphere. Sugar cane is destructive, heavily chemicalised and grown as a monoculture. When honey is the main sugar, take other sugars off the shopping list. European hive bees are the most common commercial bees but there are solitary and indigenous bees in almost every ecosystem. Recent 'humane' and more natural beekeeping systems are evolving.

Bees pollinate 1/3 of our food crops and loss of bees could cause collapse of innumerable food species. Bees are a keystone species on which an inverted pyramid of other species depends.

Needs Protection, safety, warmth, shelter from wind, rood, water, maximum solar gain, calm people.
Housing and shelter Bees have special behaviours which help them give higher yields.

It is better to grow their on-site forage than to move them around. Plant forage in groups and place hives 50m from the most intensive forage areas. This gives the nectar time to dry off on the way to the hive and prevent alcohol from being formed from the nectar.

- Swiss Bee House: Very practical set up. Open to the east, as this is the more moderate aspect.
- Kenyan Bee Hive: Good for a quickly built hive.
- Conventional Hive: Super (honey) EAST facing broodbox.

Do not site hives next to ducks or horses as bees are eaten by ducks and don't settle well around horses. Housing must have a windbreak.

Horse chestnuts and rhododendrons are poisonous to bees.

Beehives must be 1.0m above the ground so that rodents and lizards can't get into the hive.

Bee Fodder

Bees will forage within a radius of 3-5km. Scout bees report back clumps of plants so deliberately plant in clumps of about 30-50 spp. Plants don't have to be monocultures of lavender etc. but ensure there are enough plants to make it worthwhile for the bees to harvest.

Bees work over one copse for a day or more.

Apiarists and Beekeeping Associations have complete lists of wild plants, flowers and the best time for each. It is good for commercial bees to have native fodder. Bee forage requires planning. Plant species to ensure flowering all year around in warmer climates, with fewer species flowering in winter when the workers rest. Plant forage flowers on the lee side of the windbreak, especially marsh spp. and aromatic herbs.

Make a flowering calendar of bee forage with the 12 months along the bottom and species against each month.

Bees are specific pollinators of gooseberries, raspberries, sunflowers and buckwheat.

Specific honeys Citrus, linden, herb, eucalypt, orchard, blackwood, leatherwood, box, clover, bluegum etc. When one of these begins flowering, rob the hive, put in clean frames and rob the hive again when flowering has finished.

Early season flowering – Echium, willows, wattles, rosemary, native spp.
Mid-season flowering – Buddleia, brambles
Late-season flowering – Leatherwood, forest trees.
Conifers produce bad tasting honey.

Water Bees need to drink from a seepage area as they drown with steep-sided vessels. An old baking dish covered with a hessian bag will soak up water or graded banks to a dam or lake.

Bee behaviour They do not work on cloudy days and are upset by wind, so don't work bees in windy weather. When robbing the hive choose a still cloudy day before 10am.

Bees hate being moved. If they are moved then the move must be 10km or more from the base. For very short distances move them no more than 2.5cm (1in) per night.

If the keeper is feeling angry or upset, bees will pick up the emotions and can get aggressive. Some bees don't like people digging beside the hive or operating a mulcher or other noisy, smelly machines.

Bees generally swarm in spring when the new brood is raised. To collect the new brood, prepare brood boxes on poles rubbing the brood boxes with bee attractants such as lemon balm or sage. Scouts will find this and bring in the swarm. Make sure there is a queen bee excluder to stop the queen getting into the supers.

When placing a new hive, make sure that the flight path does not cross where you walk or work. It is better to have the flight path above peoples' heads.

Bee products and functions (discuss uses and economic value)

- Honey
- Wax
- Pollen
- Propolis
- Royal jelly
- Pollination of other spp
- Broods

Pollen Bees need a scraper, doormat, for their legs as they enter the hive and leave the pollen. Pollen is high in protein and used as an additive for bread which is low in protein. Bees feed it to their young.

Propolis is a silicone glue that bees make to mend their hive. It may have future commercial applications.

Wax Beeswax is a very fine product because of its quality. It is used for candles and waxing fruit boxes. Wax has a high melting point. To collect it, heat wax until melted, strain through a cheesecloth into a container with sloping edges and water in the bottom. When set it will be easy to remove.

Honey Yields Cool summers and mild winters increase yields by 30%. The value of honey per hive can be $70-80 per year, roughly $1.00 per kg profit. It is possible to rent hives out for about $30 pa. Recently rental prices have increased and so have the costs of establishing hives.

It costs about $50-60 to set up a hive unless you go to high technology which is very expensive. An integrated system has hives, brood boxes, fodder extraction and collection centre.

Stocks Bees used in Australia generally derive from Italian aggressive bee stock. Each year around about January, The University of Western Sydney has an auction of superior queen bees and they are now offering a new bee with less aggressive genes.

Some books say that native, and honey bees do not compete. Other books say that they do. Awaiting definitive information.

Student activities

- *Select an appropriate small animal for the food forest you designed last lesson*
- *Correctly site animal housing and beehives in your orchard design*
- *Design safe, healthy housing and show how it meets all criteria for safety and health*
- *Describe the multiple functions of the animal and its additional benefits*
- *Demonstrate how you will meet the optimum conditions for bees*

UNIT 19
Zone III: Cropping and large animals

Bill Mollison refers to agriculture as 'World War III against the environment' which has destroyed many ecosystems, biodiversity, drained rivers and polluted soils and water. It is essentially mining.

Permaculture Zone III is the place where we design for what we think of as commercial agriculture and so it is the site where permaculture design has most enterprise flexibility. It is a larger area dealing with crops, forage, animals and trees. This unit can only glimpse some of its possibilities.

The zone is used for all the normal stock enterprises such as dairy cows, goats, alpacas, llamas, sheep or large scale organic pig production. It is also the zone where grain staples such as wheat, barley, oats, corn, amaranth and rice are grown for home use or marketing, without adding chemicals or depleting soil or water. Also it is often adapted for increasing the size of Zone II products into commercial scale fruit, nuts and poultry or other fruit forest enterprises.

In this zone there are crops and animals which add nutrients. We design to meet the needs of the crop and to work with nature through applying restorative principles. All activities are structured within established windbreaks. Cell grazing ,pasture cropping, intercropping and mosaic patterns of enterprises are necessary for this zone. Now we have principles for regenerative agriculture and these are overriding principles.

In more tropical areas, one strategy of Zone III is alley cropping with windbreak protection. It is important that students have a mental picture how this looks. Another strategy for humid climates and small land space is Fukuoka's which was pioneered in Japan. Strategies similar to Fukuoka have been used with some success in India and Australia, and other countries.

In hotter drier climates and those experiencing global warming, the main strategy is opportunistic agriculture which requires excellent observation. The Zone is small in arid areas because these marginal lands are easily destroyed by cropping and large animals. Cool temperate climates are preferred for annual staples.

Zone III is normally larger than either Zones I and II and requires maintenance and harvesting. There are many potential crops and animals. Masanobu Fukuoka was a Japanese original researcher in zone III. People now doing interesting work in this zone are:

- USA – Joel Salatin
- France – Michel Tarrier (entomologist) – Claude Bourguignon (agronomist)
- Austria – Sepp Holzer
- Zimbabwe – Alan Savory, holistic management, cell-grazing and disturbance
- Latin America – Miguel Altieri, Agroforestry
- Tropical longterm researchers Roland Bunch and Dr Jules Pretty

Revise the criteria for placing Zone III. Mention windbreaks, establishing water in soils and dams and use of leguminous species.

Learning objectives

By the end of the unit students will be able to:

- Place Zone III accurately in a total site design
- List key factors in keeping livestock including animal preferences and suitability for enterprises
- Describe alley cropping, pasture cropping and cell grazing
- Describe how Zone III differs for different climates
- Suggest enterprises for the zone

At the end of the unit, look at an abandoned dairy pasture, or a grain monoculture and ask students to redesign it to meet the principles.

Teaching tools

- Drawings or photos of alley cropping
- Plans showing how Zone II could be expanded
- Diagrams illustrating extent of Zone III in different climates

Terms

Alley cropping Also known as 'hedgerow intercropping' is the simultaneous growing of perennial, preferably leguminous, trees, and shrubs with an arable crop. The trees are grown in 2m rows and the crop in the interspace or 'alley'.

Broadscale Farming large acreages and often limited palette of crops.

Browse Animal fodder from trees and shrubs – not grasses.

Cell grazing Keeping animals in tight groups and high stocking rates to graze an area intensively before moving to a new one.

Forage Animal fodder usually dry grasses and grains.

Grazing Animal fodder from grasses in paddocks.

HRM – Holistic Resource Management This is farm and agriculture management with attention to all inputs and outputs so they form an ecosystem. It replenishes water, builds soils and biodiversity.

Pasture cropping Similar to cell grazing with animals eating the natural grasses of the bioregion and the ground ripped afterwards to allow water and air to infiltrate.

Regenerative agriculture Agriculture which replaces more energy and nutrients than it harvests.

Staples Foods which provide energy and constitute a major part of human and increasingly animal, diets, e.g. Corn, rice, barley, oats, wheat, potatoes etc.

Ethics

Grow appropriate staple crops first for home use, then for the bioregion.

Build natural resources such as soil, water and plants to achieve yields and leave more than harvested.

1. Cows and droppings on pasture

2. Chickens scratching and enritching

3. Pastures recover and grow well

4. Cows grazing again

Cell Grazing

Principles

The following principles are derived from several sources and mostly from ILEA – Intensive Low input Ecological Agriculture.

- Put up the set of principles and ask students to think of a modern wheat field and what is happening that is counter to these. Or find a good photo of a huge grain monoculture harvest and ask what would have to change
- The first most important principle is that no enterprise ever removes more nutrient, water and energy from the land than it puts back

Other principles follow:

- Nothing is wasted
- Water, soil and air are not polluted
- Resources such as energy, water, species are used at a rate below which natural cycles of the Earth can replenish them
- All life; large/small, human/nonhuman is granted the intrinsic right to wellbeing

Resilience and the principles of permanence

Resilience is now accepted as a necessary goal of agriculture and the following principles help achieve it:

- No energy subsidies or extra nutrient sources – only solar energy and internal nutrients
- All wastes assimilated – no toxins and no pollutants
- Net energy yields – human and animal labour
- Humans: 18 calorie return for 1 cal input
- Machines: 1 calorie return for 5 cal inputs
- Odum – net energy is the only energy with true value to society and its yields must endure through time.
- Use plants life cycle successions and use all products – maintenance not output.
- Energy spread evenly throughout the whole system – uneven outside energy input removes local control.
- Resources such as habitat are productive capital and to be preserved for the future and never run down.
- Use polycultures and diversity – plant spp. and cultivars, diversity, maturity and resistance to perturbation
- No enterprise is started before its needs such as water, protection and nutrient are physically established

Characteristics of Zone III

- Connected to Zones I and II for easy access
- Hardy trees, bush species, ungrafted, and forage species
- Water systems developed through out

- Windbreaks and firebreaks
- Spot or rough mulching and young trees protected
- Self-forage systems, and browse trees; herbal pasture
- Nut tree forests
- Sustainable and soil building
- Contains wildlife
- Animals – Sheep, goats, dairy cows, bees, geese, pigs
- Broadscale forage sytems – Fukuoka style – well-rotated

Ask students to site Zone III in their designs and choose enterprises.

Siting Zone III

Zone III can be continuous with Zone II or directly lined up with Zone I. It depends on the nature of the land and the farmer's wishes. It is usually connected to Zone IV and sometimes Zone V. In the last case there may need to be a filter or buffer of plants to remove nutrients and plants which might encroach into Zone V.

- Keyline the site
- Choose flatter land below keypoints which can also be used for water holding points
- Ensure water availability by gravity from high dams along swales and contours
- Dams can also be in windbreaks
- From sector analysis site extensive windbreaks which also function as low maintenance orchards and timber sources, wildlife shelters and firebreaks
- Identify good access to farm buildings, with doors opening to the morning sun
- Agroplough the land if necessary

Alley Cropping in Zone III

Tree Species

The system can succeed or fail on tree selection.

Criteria for tree choice:

- Ease of establishment from seed or cuttings
- Rapid growth rate
- Good coppicing potential
- High N-fixing capacity
- Deep-rootedness with roots at different level from crop
- Multiple uses e.g. Forage, firewood, erosion control
- Ability to withstand stress e.g. Drought, pH and fire
- High leaf:stem ratio
- Small leaves or leaflets
- Freedom from pests and diseases

Species successfully used in tropical/sub-tropical areas:

- Legumes: *Leucaena leucocephala, Gliriciaia sepium, Cassia siamea*
- *Erythrina* spp., *Tephrosia candida, Sesbania grandiflora*
- Non-legumes: *Acioa barterii, Alchornea cordifolia, Gmelina arborea*

Species with potential under Australian conditions:

- *Sesbania sesban, Calliandra calothyrsus, Albizia chinensis, Albizia lebbek, Acacia fibriata, A.cunninghamii, A.cinncinata*

Temperate regions:

- Tagasaste, *Chamaecytisus proliferus*, mulberry, poplar etc

Tree Row Spacing for Alley Cropping As the row width increases the available mulch per unit area of crop decreases and the effectiveness of the mulch to supply nutrient may also be reduced. In close rows, a greater amount of mulch will be available but the number of crop rows will be reduced.

A compromise must be reached. Climate and soil type must be considered. Spacings commonly used are from 2-5 metres.

Tree Row Management Is the timing of first cut, cutting height, and cutting frequency for the productivity and effectiveness of the row trees.

Stem and root development depend on the time between planting and the first cut. A well-established tree has more reserves. Generally, the establishment time is between 6-12 months before the first alley crop is planted.

A low cutting height is desirable since it reduces shading effect of the trees on the crop. Cutting frequency is adjusted first to the growth rate of the tree and secondly to the growth of the crop. In general, two to three cuts per growing season of six months seems to be the best compromise.

Applying Mulch Before and during the crop season, trees are pruned for mulch with a mulching machine. Mulch improves soil organic matter and provides nutrients, particularly nitrogen to the crops.

With a mulching machine, mulch can be left on the surface, or incorporated into the soil before crop establishment.

Surface mulching helps to control weeds and assists water retention in soils but experimental evidence suggests that incorporation improves the efficiency of nutrient transfer to the crops, e.g. Only 38% of nitrogen in Leucaena spp. prunings was recovered by maize but when incorporated into the soil, recovery increased to 63%. Leaves of nitrogen fixing species are higher in nitrogen than in non-nitrogen fixing species.

Crop Species Within Tree Row Crop selection is less important than the tree selection and management. Most cereals appear to adapt to alley cropping and the system is flexible enough to allow for changes for specific crops.

Maize has had very good success, sorghum, cassava, upland rice and pineapples have also been successfully used.

Benefits of Alley Cropping

- Stabilisation or increase in crop yields due to the addition of nutrients and organic matter to the soil/plant system
- Reduction or elimination in use of chemical fertilizers
- Trees aid nutrient recycling by exploiting moisture and nutrients deep in the soil out of reach of most crop roots
- Physical improvement in soil because trees modify soil temperature fluctuations and reduce soil moisture losses. The additional organic matter from the mulch improves soil structure resulting in better water infiltration, and less run-off
- Tree rows on sloping land act as physical barriers to soil and water movement with less erosion. Extra products from the systems such as forage, firewood or timber
- Weed control by firstly shading during early vigorous growth, and later from mulch

Disadvantages of Alley Cropping

- Possible competition between trees and crops for light, water and nutrients, especially in the early phases of development
- Reduction in crop area due to tree rows
- Additional labour required for establishment and pruning of tree rows as well as mulch spreading. Some limitations on the flexibility of land use due to presence of woody plants

Overall, the biological merits of alley cropping appear to outweigh the disadvantages.

Opportunistic agriculture

This is agriculture in which farmers wait for the right conditions for planting their crops. It is important in dry areas and where climate is becoming increasingly unreliable. So farmers have to meet the needs of the land for a crop such as sorghum with water harvesting (but not full irrigation) windbreaks, deep ripping and so on. Now they wait for the rains sufficient to germinate the seed and sprout it. They will use their water strategically to obtain a yield. They can also do this with fruit and nut crops where water is applied from dams etc. as biologically required for flowering or fruit set – again to obtain a yield. It requires good observation of local conditions.

Nut Trees in Zone III

Nut trees are a good enterprise choice for Zone III if staples or animals are not required. Best grown as low maintenance orchards with plantings of the following:

Almonds Warm, dry summers, lighter soils, two trees. Budded to a peach or almond seedling stock fairly easily. Plant at 6-7m apart. Bear at five years, yield/tree about 4kg. Harvest Feb-May, Western NSW.

Pistachio Long, hot dry summers, cool winters, tolerate wide range of alkaline and saline soils but like good drainage. Need one male pollinator to every nine female trees. Plant 9m apart. Bear at five years and tend to crop heavily every second year. Start with 1.0 kg/annum and increase to 30kg/annum. Harvest in March. Dehull for keeping well.

Chestnuts Cool, mild climates, deep soils, bud or graft. Grafted trees bear to 90kg at 5-7 years. Plant 12x12m.

Walnuts Cool to warm climate, deep soil, good drainage. Bud or graft. Plant 15x15m. Bear at 10 years, in March, 25-50kg.

Pecans Temperate to sub-tropical climates with well-drained soils. Propagate through budding or grafting; some are self-pollinated. Plant 12x12m. Yield in May to 50kg at maturity.

Hazelnuts Prefer cool to cold climate with high rainfall. Propagate from layering or rooted suckers. Pollination – several varieties. Harvest Feb-Mar. Bear at 5-6 years at 4kg.

Bunyas

Macadamias or Hawaiian Nut

Animals in Zone III

Animals are essential to balance this ecosystem. They may be the main enterprise or support another one. Animals are naturally a part of a system, not a means to an end. Animals are great waste converters, lawn mowers, pruners, heat sources, methane producers and friends.

Modern systems are often wasteful, cruel and extremely inefficient. Industrial agriculture feeds high quality food to animals, and forests are cleared to feed them. Meat is a poorer quality product. Machines, used in place of animals are degenerative and lose value while animals increase in value.

Always prepare their environment before buying or putting animals into a system.

When electric fencing is used, dairy cattle, pigs, and goats can be grazed in alleys. Animals should only be on area for a short time to reduce compaction.

Ruminants

Cell grazing is the primary animal management strategy in Zone III. Stocking rates are critical and animals are managed by observation rather than prescription.

Grow animal feeds according to site and climate.

Animals grow well and have greater disease resistance when they have 75% grasses and 25% browse – shrub and tree leaves and fruits.

Pasture Crops Rye grass, prairie grass, K.V. rye, fescues, clovers, medics, lucerne, brassicas e.g. Rape, kale, turnips.

Holistic or Mob Grazing In Australia, the value of indigenous grasses is becoming recognized and animals are grazed on mixed pastures intensively so they eat all grasses instead of selecting and this means they eat weeds and medicinal plants. Their health improves. Once the sward is grazed down the animals are rapidly moved on to new pasture which again they eat down. In this way they dump a large nutrient load on the grazed area and don't return to that place for up to four months and so it recovers extremely well. In this way the savings of industrial fertiliser, and drenching animals is reduced by $40,000 per year for a flock of 2,000.

Tree Crops Tropical and Sub-tropical – Paw-paw, yams, bananas, palms, guava, etc.

Temperate Persimmons, mulberries, apples, acorns, bulrushes, olives, chestnuts, honey locust, loquats, figs, carob.

Weed Crops Pigs eat couch, bladey grass, kikuyu, oxalis, buffalo grass, bracken.

Goats eat brambles, rose cuttings, lemongrass, comfrey, pampas grass, blackberry.

Cows eat willows, poplars, tree lucerne, carob and others.

Keeping pigs

Pigs are originally marsh and forest foragers. Their advantages are: Very efficient: waste convertors, hoes/Tractors lawn mowers, weed controllers.

They function at two levels called tractoring.

The first is a high stocking rate where they take everything from the ground and eat it and turn over the soil.

The second is a lower stocking rate called maintenance where pigs graze in orchards or food forests, cleaning up diseased fruit and adding fertiliser to the trees.

Pig Tractors Where they clean up land of weeds before it is used as orchards, market gardens and grain crops. At this intensity of stocking rates their efficiency is different.

100 pigs on five acres will completely tractor it in six weeks, cleaning up tiger snakes, bracken, weeds in Australia. 100 pigs on 100 acres take 18 months.

20 pigs will tractor l acre.

1 pig will plough 100sq.m in six weeks.

(Tractoring rates will vary from climate to climate.)

Housing Single pen.

Show pictures of pig housing

In a large pig system, keep male and female separate.

Sloping floor – pigs like cleanliness. Collect dung and grow water hyacinth as a good pig feed and frost will kill it in winter.

Water Pigs eat 10.8kg (24lb) of wet food per day and need 45.5l (10 gals) of water (in Australia and similar climates). They need a pond, or mud patch because they don't sweat.

Food They eat all the foods listed above and also have strong teeth for grinding nuts of all types. Grow all pig food, or feed household scraps, dairy excess, windfall fruit. Grow pig food in alleys – vegetables, fruit and grains. Don't buy it.

Pigs are often iron deficient so feed them comfrey, chicory and parsley.

Keeping dairy cattle

Cattle are descended from forest foragers of India.

Siting Design for single shed close to where animals can be milked once or twice a day. Usually on a higher place where accumulated nutrient moves downhill by gravity.

Design for larger dairy cow unit. As for all animals, rotation is essential for controlling disease and effective grazing.

Housing Cattle do not need housing so much as shelter. Carefully design shelter belts to avoid very hot or cold conditions and for calving, (a bracken patch is good). Lack of shelter in hot or cold conditions can reduce production by 15%.

Water is very important for high production. Generally take the water to the animals because hooves are extremely damaging on the edge of dams, rivers and lagoons.

If there are soil deficiencies add nutrients to drinking water and cows will urinate it back into the soil.

If there is heavy rain after a dry period, many crops are toxic with high cyanide and alkaloid levels. Keep animals off new growth for 3-4 weeks.

Food Cattle will eat all animal foods mentioned in this unit, but must have sugar to digest roughage. If fewer than 15 cattle, it is better to keep them penned and take cut grass to them. Cell grazing is valuable as well.

Cattle eating naturally consume total feed as 25% browsing and 75% grazing. They can only eat 7-8 hours per day because they are ruminants and need time to digest. They prefer longer grasses – 11cm (4-5in), when grazing.

When rotating, graze cattle before sheep, as sheep graze shorter and closer to the ground.

Alley cropping is useful for larger herds. Animals are rotated which is important for ground compaction recovery and disease control. Use a Wallace plough (an agroplough) regularly to aerate the soil.

Animals need a consistent supply of feed so draw up a calendar to ensure seasonal supply. Tree forage can raise stocking rates.

In dry periods animals need a urea/water molasses lick.

22% of the pasture area turned into a well designed windbreak can double the yield of the remaining 78% due to edge effects and pest control.

Cattle do very well on willow, poplars and tree lucerne. e.g. Willows give six tonnes of feed per ha. Tree lucerne 4t/ha. (grains are much more work and money to get these quantities).

Make sure cattle come to you. Always call them with a food incentive such as a handful of grain.

Keeping goats

In general, do not keep goats because they are the final element in ecological destruction. If you must, then graze them on boxthorn and blackberries.

Siting Goats need very good shelter on dry land, and protection from wind, rain and cold. They are mountain animals, not pasture animals.

Housing Use ringlock fencing, sleepers two feet into the soil and six feet above with two rows of barbed wire at the top. Use wire strainers as they will climb up timber ones.

Water Never put them on wetlands and, take food and water to them. In SE Asia goats are kept off the ground in bamboo cages.

Food Goats need much roughage and will eat brambles, rose cuttings, lemongrass, acacias, fodder trees, lemonbalm, comfrey, pampas grass, mulberry, prunings...

Advantages Excellent milk, Great manure, Good hair, Good weed controllers.

Disadvantages Cunning and destructive so fencing is all important.

Breeds Milk – Anglo Nubian (rich, creamy), Sarinham, Alpine (high in cream), Togenburn. Anglo Nubian are hardy, take heat, and are good for fetta cheese. Saanen produce bulk milk and angora for fleece.

Sheep

Siting Sheeps are creatures of habit so place suntraps and shelter belts where animals like to graze and rest. Build a central pen with mesh fencing.

Water Carry to animals via keyline or dams or bores.

Food Eat most things.

Advantages Excellent for wool. Good meat. Milk for cheese.

Disadvantages They attract flies, although if kept clean around the crotch and well cared for they remain healthy.

Horses

Discuss place, climate and food needs with local advantages and disadvantages.

Don't feed Leucaena to horses because it contains mimosine which is toxic to horses.

Generally Incompatible Animals

- Chooks with cattle or pigs for the risk of TB in cattle and pigs
- Chooks with turkeys because both get mites and blackleg
- Chooks with sheep can have Salmonella problems
- Chooks with goats because Coccydiosis carried by chooks is lethal to goats

Animals need protection within 200 metres against cold winds.

- Chooks with ducks because ducks foul water
- Goats with horses as horses susceptibility to tetanus increases.
- Goats with sheep – parasites proliferate, especially Barbers pole worm which breeds in sheep
- Ducks with bees – ducks eat bees
- Bees with horses – horses go stupid

Student activities

- *Completely redesign a monocultural maize field to be diverse, resilient and abundant*
- *Design a Zone III for their property and suggest appropriate enterprises*
- *Describe the positive and negative effects of keeping livestock in this area*
- *Describe briefly your reasons for an alley crop design and include any disadvantages of the trees used. Note at least two major characteristics of Zone III for climates different from your own.*

Case Study

Results of an exercise called Converting the Rice Field in a PDC in Viet Nam

While talking about the rice monocultures in this part of the world, I asked the course participants four questions:

1. How can you increase the number of species growing in the ricefield and along the bunds?
2. How can you grow more trees in the rice fields?
3. After the rice is harvested, what can you grow in the fields?
4. If you decide not to grow rice what can you grow instead?

Participants found these ideas very strange and discussed them at length. They then made drawings of possibilities.

Principles: Sustainability, Permanence and Resilience

Craig Linn – Sustainability

Nothing is wasted. Water, soil and air not polluted. Resources such as energy and water species are used at a rate below which natural cycles of the Earth can replenish them. All life; large/small, human/nonhuman is granted intrinsic right to well-being. People to live within their means with respect for the universe.

Resilience and the principles of permanence

- No energy subsidies or extra nutrient sources – only solar energy and internal nutrients
- All wastes assimilated – no toxins and no pollutants
- Net energy yields – human and animal labour
- Humans: 18 calorie return for 1 cal input
- Machines: 1 calorie return for 5 cal inputs
- Odum – net energy is the only energy with true value to society and its yields must endure through time
- Use plants life cycle successions and use all products – maintenance not output
- Energy spread evenly throughout the whole system – uneven outside energy input removes local control.
- Resource seg. habitat are productive capital and to be preserved for the future and never run down
- Use polycultures and diversity – plant spp. cv diversity, maturity and resistance to perturbation – layers of garden and wild garden species

Human behaviour for resilience and permanence

- Accept a lower material standard of living for higher quality of life. e.g. Unpolluted food, air and water
- Slower rate of technological change
- Decentralise power and production – from small communities out
- Limit human population growth and spread
- Move to diversified agriculture/organic gardening, polycultures of life and let live
- Create and maintain a varied environment i.e. balanced mosaic

Roland Bunch – Five Principles of Sustainable Agriculture for the humid tropics

- Maximise organic matter production
- Keep the soil covered
- Zero tillage
- Maintain biological diversity
- Feed plants through the mulch

Chad Helwinkel and Daniel de la Torre Ugarte

WINNER, GLOBAL ENERGY SECURITY CATEGORY for the Farm Foundation essay – transition to regenerative practices that:

- Sponsor their own energy
- Build soils
- Produce in abundance.

These three policy imperatives should be the long-term guideposts in setting all policies that affect agriculture.

Barry Traill: Sustainable agriculture

There is now strong evidence that regenerative and resource conserving technologies and practices can bring both environmental and economic benefits for farmers, communities and nations. The best evidence comes from countries which have been largely untouched by modern external technologies such as pesticides, fertilisers, and machinery. In these complex and remote lands, regenerative technologies have substantially improved agricultural yields often using few or no external inputs.

All successes had three elements in common

1. All have made use of locally adapted resource-conserving technologies
2. Secondly, in all there has been a coordinated action by groups or communities at the local level e.g. Land Care?
3. Thirdly, there have been supportive external – or non-local government or NGOs working in partnership with farmers.

In analysing a site, monitor and recommend:

1. More thorough incorporation of natural processes such as nutrient cycling, nitrogen fixation and pest/predator relationship into agricultural production processes.
2. Reduce those off-farm external and non-renewable inputs with the greatest potential to harm the environment and or health of farmers and consumers – and minimise variable costs.
3. More equitable access to productive resources and opportunities.
4. Greater use of biological resources and genetic potential of plant and animal spp.
5. Greater productive use of local knowledge and practices including innovative approaches not yet fully understood by scientists or widely adopted by farmers.
6. Increase in self-reliance among farmers and rural people.
7. Improvement in the match between cropping patterns and productive potential or constraints of climate and landscape to ensure long-term sustainability of current production levels.
8. Profitable and efficient production with an emphasis on integrated farm resources and conservation of soil, water energy and biological resources.

Barry Traill Publisher ANU Press 2007

UNIT 20
Zone IV: Restorative forests

This zone is continuous with Zone III and sometimes with Zones II and V. It is a carefully designed forest belt providing a wide range of high and low quality tree crops. It is permanent and selectively harvested so as not to destroy the forest structure.

Many trees on a farm function as a forest if the canopy closure is from 50-70% and they perform the same forest functions of increasing soil moisture, reducing erosion, river silting and flooding, preventing desiccation, improving rainfall and soil, and producing mulch. They are harvested for their timber and non- timber-products (NTPs); essentially the 7Fs: Fuel, food, fodder, fruit, flowers, fibre, fertiliser.

Trees as the anchors of soil and providers of clean water can be considered the basis of every civilisation. They are fundamental to permaculture as permanent culture. Ideas for this unit were presented in units on Ecology and Forests.

The most important thing we can do is to plant trees for our children to protect the land and its waters. Students need to calculate how much timber they have used in their lifetime – include buildings, firewood, paper etc. and then calculate how many trees they need to plant each year so their children and grand-children can have the same amount of forest products. This Unit helps us know how to replace all that we have taken. They are called restorative forests.

Students have selected and placed zones from I to III on a plan. Zone IV usually lies between zones III and V on farmland. In towns, perhaps there is only the possibility of a few small perennial shrubs or wildlife corridor along back fences. In some cases schools and hospitals can have these extensive forests which will be valued for centuries. Some old cities had a circle of these forests around the city.

The Tree's Prayer

Ye who would pass by and raise
your hand against me,
hearken ere you harm me.

I am the heat of your hearth on
the cold winter nights, the friendly
shade screening you from the summer sun;
And my fruits are refreshing
drafts quenching your thirst as
you journey on.

I am the beam that holds your house,
the board of your table,
the bed on which you lie,
and the timber that builds your boat.

I am the handle of your hoe,
and the door of your homestead,
the wood of your cradle
and the shell of your coffin.

I am the gift of god
and friend of man.

Ye who pass by, listen to my prayer – Harm me not.
From a Tree Nursery in Kenya.

Learning objectives

When they complete this unit students will be able to:

- Describe the characteristics of Zone IV
- Select trees correctly for home and commercial products
- Establish trees with maximum success
- Introduce appropriate animals

Teaching tools

- Briefly revise Forests
- Ask a student to draw the succession graph from Ecology Unit on the board
- Prepare drawings for two types of forest; one with animals, and the other without animals

Terms

Coppicing Cutting a tree or shrub so that it sends out several new stems and continues to grow and can be harvested.

Canopy closure The degree to which the leaves of trees meet and close out light. It is usually expressed as a percentage of shade e.g. 50% or 70% canopy closure.

Filtered light The way light moves through a leaf canopy. Conifers have 'black' shade permitting few things to grow underneath them. Eucalypts have 'filtered' light allowing a rich and diverse sub-storey to grow under its canopy.

Pelleting Coating seeds in mud, peat moss etc. and then scattering them and so the seed is protected from being eaten by ants or birds, and also from dessication.

Woodlots Land set aside specifically for growing trees for harvest.

Ethics

Plant trees consciously and responsibly to have future sustainable timber supplies and to assist in modifying climate and cleaning water.

Principles

- Take a long-term approach by consciously planning for future harvesting of whole trees and also for their products without cutting the trees.
- 40% of every site should be tree covered just as 15% of every temperate and wet tropical farm should be water
- Each tree's harvested products should last as long as it took the tree to grow – this eliminates woodchip for newspaper
- Encourage students to think long-term and to help their clients to think long-term
- E.g. Until recently French farmers in Normandy, at the birth of a child, used to plant sufficient trees to provide for the house and furniture of that child at their marriage. People need to think of trees and their products for their grandchildren, while at the same time many of the products increase the individual's and the land's sustainability. i.e. self-sufficiency, economic viability and real wealth

Characteristics of Zone IV

- Long term permanence and structure which harvesting never destroys.
- Grow into contour shelterbelts.
- Provide for firewood, building, precious timbers, honey, fodder.
- Minimal watering preferably from sewerage or grey water.
- Hardy range animals such as wool sheep, beef cattle, emus, deer.
- Animals maintain the forest with minimal feeding, and at appropriate stocking rates.
- Products increase in value with time

Benefits of this forest zone

Ask class for ideas and elaborate on the following:

- Control of soil erosion on slopes, water courses and ridges
- Preservation of wildlife especially birds and insects for pollination
- Gains in farm yields e.g. Animal products, orchards and cash crops
- Forage needs of animals in hard times
- Provision of firewood and building materials
- Supplementary products when crops fail due to adverse conditions
- Stabilise climate and buffer against excesses of cold, heat, wind
- Assured storage, income and marketing of many tree crops

Range of forest yields

Elicit answers from class – use a brainstorm to expand the yields especially those which require further processing. Try for about 100 yields from forests. Be clear that there are Timber and NTPs – Non Timber Products and these latter are enormously important – often for survival in harsh times.

- Fungi
- Seeds
- Nuts
- Coppicing
- Timber
- Small fruit
- Wild fruit
- Honey
- Poles
- Bark (dyes, craft, tannin, medicine, mulch)
- Firewood

Design priorities

Site selection

1. Base the site on available water for establishing trees. Revise Yeoman's water harvesting design. Select land which may be sloping but not greater than 45°.
2. Design carefully with the main restraints usually being labour and machinery, then slope, drainage, soils and the requirement for self-reliance.
3. Design a windbreak for fast, effective establishment of the final crop seedlings.
4. Zone IV must include plants which function as speciality crops, conservation, wildlife corridors, hedgerows, bee forage and woodlots.
5. At maturity this Zone IV grazes domestic or indigenous animals.

Begin planting using the principles of succession for greatest success

1. Start with nitrogen fixing species to enrich the soil
2. Interplant throughout this first flush forest with the climax species

3. Harvest the nitrogen fixing species for firewood or mulch etc
4. Harvest productive trees each year and never clear fell
5. Replant every year and increase planting to rebuild forests

Types of tree farms

Farm type 1: Harvest tree crops and domestic animals

When this tree system is mature the effect is one of lightly timbered woodland with filtered light. Animals graze grasses and clovers which thrive in filtered light and they also browse (up to 40% of their intake) fodder trees and tree fruits.

Setting up

- 12 to 16 hectares is minimal for this farm type
- The forest edge or windbreak is critical and needs at least 600m depth
- Choose the mature tree layout by frequency of harvesting i.e. Intensity of care and work
- Timber crops are selected as high-value trees widely spaced in rows but more closely spaced in columns to allow good pasture development in between rows for grazing
- Start planting the whole site in the first year with short term nitrogen species such as, *Acacia decurrens, A. dealbata, A. melanoxylon*, Tagasaste (splits easily), Casuarina (hot burning firewood) and clovers which like filtered light. Kurrajong and Brachychiton are good firewood and fodder trees
- As these mature, plant the long term timber and speciality trees
- Animals are let in as the trees mature (years 3-6) and grazing is controlled. Leaves of poplars are higher in protein than lucerne.
- Some fodder trees are planted initially but others are added progressively to give about 250 trees per acre worth about $2,000 per tree in 13-14 years
- Planting Methods, (See Zone V – Bush Regeneration) for ripping, pelleting, up-wind seeding and planting

Farm type 2: Harvest only tree crops

From the beginning of time, forests have been the refuge and survival of people when bad times struck. Now they are badly diminished and today many traditional peoples are severely impoverished because of their loss.

This tree farm grows trees for timber and also trees for their non-timber-products (NTPs). These are all the products from forests which do not require the trees to be cut down. They supply hundreds and perhaps thousand more products than simply timber.

Do a quick brainstorm of these.

Indigenous or wild animals are allowed to live in this forest unless they are particularly destructive and carry out all the functions the forest requires such as pollination, nutrient supply, cultivation, pest management and so on.

- 50 or more hectares is required for a commercial forest and can include fruit crops
- Start with a 5-row windbreak and coppice one row every two years. Coppice products will be mulch/fodders, for example, willows yield six tons/ha whereas lucerne yields four tonnes/hectare
- Woodlot farms vary in design and products. It is possible to plant five types of tree products. They are often mixed

1. Firewood production species have a high calorific value and coppice well. Harvest these from year two with 1/7th being harvested annually either as stockwood or logs 4-10cm diameter i.e. stove fire-box size. High calorific value trees: Acacia, tagasaste, casuarina, yellow box, red gum, also chosen for persistent coppicing whereas eucalypts, ironwood, bloodwood, angophora, etc. are chosen for self-pruning qualities

2. Polewood production species are two types depending on their uses:

Very durable, whose uses are fencing, house and furniture construction e.g. Chestnut, raspberry jam acacia (*A. acuminata*), black locust, cedars, some eucalypts e.g. Turpentine, red river gum.

Less durable, whose uses are scaffolding, formwork, chipping, fuel bricks, fibre, cellulose, stockfeed, oils e.g. Poplars, willows, acacias, Chinese elm.

3. Long term fine timbers These add to the capital value of the farm from year one. Many have complementary pioneer species inter-planted for medium term yields e.g. leguminous trees and small cedars.

Their greatest value is generally from 40-100 years for fine furniture, inlay, panelling and plywoods.

Some species:

- native red cedar ($1600/m^3)
- silky oak
- paulownia
- jacaranda
- blackbean
- red bean

- cudgeria
- white cedar
- Burmese rosewood
- rain tree
- Honduras mahogony
- Spanish mahogony
- swamp cypress
- mulberry
- blackwalnut
- rosewood
- redwoods
- black ebony (sells at $28,000/m^3)
- oaks (three species)
- fine cedars
- blackwood
- white beech
- Brazilian pepper tree
- albizia (two species)
- camphor laurel
- Queensland maple
- tulipwood
- Transvaal teak

4. Hedgerow and contour forests have trees planted in hedgerows, contour bank forests, roadsides, watersheds and steep slopes. Plant between six and 30 species for fruits, nuts, forages, special wildlife habitat, browse, mulch and stockwood. Gather fallen products and constantly replant.

5. Special forests are those planted for special ecosystems or products, for example, for wetlands or, acid uplands Plant species to enhance other farm enterprises or to supply local needs.

Many crafts are disappearing as the timbers used for them are disappearing from wild sources. Fewer mill- hands know about them and yet hobby people are setting up new demands. Some examples:

Rattan – A climbing palm for furniture in global acute shortage. There are hundreds of species which have a low environmental impact and, residual areas, such as: Borneo, Malayasia and the Philippines, are being logged out.

Bamboo – For reinforcing, furniture, fences, irrigation pipes, scaffolding, pergolas, baskets, fishing poles, stakes, downpipes is in considerable demand now as flooring and walls in Japan and California.

Tea-tree – uses are mainly oil, fencing, bark, wallpaper.

Barks – as spices or medicines are cinnamon, quinine, quassia, slippery elm.

Basketry species are traditionally willows, hazels.

Student activities

- *Add this zone to their own site design. Complicate matters for them by asking what they would do in an arid zone – or tropical zone*
- *Ask them how they will water their trees. Remember that such forests are best watered from community sewerage or grey water*
- *Ask for details of the edge effect that will border with Zone V*
- *Describe the timing of the establishment of their forest*
- *What is the most suitable animal to use with trees selected for Zone IV for their region?*

UNIT 21
Zone V: Natural forests

Natural or wild forests are old growth, or regenerated forests. They are inalienable and maintain the good health of the land. They protect soil, take up water, conserve indigenous species and their genetic stock. They can mutate with global warming yet assist climate stability and reduce risks on site. It is the protective forest. They play a large role in maintaining climate stability. Everywhere indigenous forests are being removed very fast. It is urgent to save all remaining natural forests and old regrowth forests. They should never by felled.

This is the final zone in the permaculture design method and it protects boundaries and waterways, roads and ridges. It is normally continuous with Zone IV but may merge through windbreaks with Zone III or even Zone II. It should cover all difficult sites such as wetlands, marginal soils and landforms with unique ecosystems. The unit on Forests provides the concepts for this unit.

Once students have sited Zones 0, I and II on their land they should place Zone V next because it will protect Zones IV and III and help place them most advantageously. Many forests have weed and feral animal infestations so students must know how to assess the health of a forest. They need techniques to analyse forests damage to see where it originates e.g. Weeds, water, pollution, and what remedies are possible. Management of natural resources is closely linked with how we view and understand ecosystem structure and function.

Students must know what the restoration priorities are for these forests and how to start. They must be able to place this zone on their designs.

Natural conservation forests can be established from degenerated land but it is almost impossible to replace natural forests completely and exactly since some species we do not know may have already become extinct.

Species *Refugia* are remnant species from previous, very old continents and cultures. For example, in Western Australia there are more species per hectare on arid lands than there are in tropical rainforests and many species belong to stock strains that have been extinct for centuries elsewhere. Australia returns strains back to other countries and continents. For example, the Sahara Desert has only five desert species left in some places and Australia is providing replacement stock.

Die back, like salination of soils, is such a serious and wide-ranging phenomenon of rural land that it is worthwhile running through the causes, i.e. grain and grazing, high nitrogen levels, proliferation of scarab beetles on pasture, build up of tree skeletonising pests, lack of understory (like *Bursaria* spp.), loss of bird and insect predators, cool fire burns leading to increase in *Phytophthera cinnamomum* and removal of major tree cover.

There are two situations for extending these forests:

- The first is re-establishment from bare damaged land
- The other is extending remnant patches of vegetation

Learning objectives

Students will learn to:

- Place this zone on a design
- Identify two main types of conservation forest
- Develop a forest from remnant vegetation
- Encourage wild native animals and plants
- Control feral animals and plant pests
- Implement the main procedures in bush regeneration

Teaching tools

- Photos, slides, posters of natural forests
- Drawings showing an analysis of weed infested site
- Sketches of stages in restoring forest starting with least infested
- DVD *On the Edge of the Forest* with F. Schumaker
- A walk in a forest or give the class in a forest or remnant vegetation to do a weed analysis

Terms

Alien A plant or animal foreign to a local bioregional.
ANR Assisted Natural Regeneration
Aquifer A subsoil water lens.
Endemic Occurring naturally in a local area.
Escape A naturalised plant, exotic, which moves into natural ecosystems.
Extinction The complete wiping out, annihilation, destruction of a race, species etc. The end of millennia of evolution.
Exotic Of foreign origin, not native, introduced from abroad. Not fully acclimatised.
Feral Animals and plants that have run wild, or returned to a native state.
Indigenous Occurring naturally in a land.
Introduced Introduced from overseas. In Australia, a plant that was not present before European invasion in 1788.

MDTs Minimal Disturbance Techniques.
Native Belonging to a place by right of origin.
Naturalised Introduced into a region and flourishes as if a native and implies the plants are weedy or invasive.
Wildlife corridor Not a row of trees but a piece of natural forest, or designed pathway along which animals and plants can migrate, or seek sanctuary when under threat.

- *Ask students what this zone provides and especially how it will function in their design. Revise Forests here, and include recharging aquifers and conserving slope stability as special functions*
- *Ask the importance of conserving local indigenous genetic stock*

Ethics

Preserve and care for all remaining forests and establish wildlife corridors.

Extend and plant these forests greatly everywhere for intergenerational equity and climate stability.

There is now a branch of legal studies called wild justice or wild law occupied with the rights of wilderness such as forests, trees and rivers. This is exciting.

Principles

- Preserve all fragments of natural forests
- Place Zone V along all rivers, waterways, farm boundaries, all ridges and slopes greater than 45 degree slope and cross and link the land with wildlife corridors
- Encircle all human settlements with conservation forests
- In permaculture designs make Zone V as large as possible

Forest management problems

Overuse along fences, dams, under single trees

Incompatible use by horses, motorbikes, trail bikes, 4WDs

Increased water volume from houses, dam spillways etc.

Erosion from water, wind, compaction

Water pollution from fertilisers and herbicides, from roads, gardens, crops

Air pollution from factories, cars, detergent breakdown

Rubbish Non-compostable such as plastics, bio-accumulative toxins

Fire Hazard reduction burns, prescribed burns

Feral animals Carnivores/herbivores, dispersed seed, eat to extinction

Decreased native animals, loss of pollinators, maintainers.

1. Re-establishing conservation zones along rivers, hills and roads

- Start by protecting the regeneration site such as rivers, hills and roads
- Select and plant indigenous pioneer species
- Control pests such as rabbits which destroy new plants
- Take great care with the edge which must not be less than 600m
- Keep all exotic stock off the regenerating land
- One old tree can have sufficient seed, if fenced off, to start a new area
- Encourage birds to deposit seeds via faeces by putting stakes in the ground to act as perches
- Weed carefully as plants start to grow
- Conservation zones will always need management. In Australia, Echidnas are a sign of a healthy habitat and bushland

2. Extending remnant vegetation into forests

Size and shape Success in maintaining remnant forests in good condition depends partly on the size and shape of the land. So, for example, a 2ha site will have about 0.7ha or 33% undisturbed. A 64ha reserve will have about 54.8ha or 86% undisturbed.

The shape of the reserve is important. The closer the shape of the site to a circle the less the invasion of weeds and pests and the more vegetation will be in tact.

Weed invasion is the major natural problem in keeping conservation zones in a natural state.

Weed site analysis Map the weed invasion for correct remedial measures e.g. If the problem is overgrazing, it is little use to spray weeds.

Are the weeds primarily: annual, biennial or perennial?

Classify weed density as:
1. Weedfree forest
2. Sparse weeds
3. 50/50 weeds:native
4. Dense weeds
5. Total weed

Map these areas by marking them with these numbers on map.

Preserve all sacred and indigenous trees

Layers of weed infestation are structural:

1. Weeds mostly in understorey e.g. Creepers like Tradescantia spp.
2. Weeds intermingled with native shrubs e.g. Mickey mouse plant
3. Weeds in overstorey e.g. Camphor laurel, privet

Weed control and removal strategies

Discuss advantages and disadvantages of each:

Removal of plants or seed heads

- Annual – remove before seeding
- Biennial – remove before flowering
- Perennial – remove when seedling
- Change growing conditions
- Make soils wetter, drier, shadier etc

Large scale mechanical clearing (for establishment of new areas)

- Mowing, especially for annuals/plants with many seeds
- Herbicides
- Burning
- Shade, weed control mats, or dense planting for shade Minimal Disturbance Techniques (MDT)

Minimal Disturbance Technique (MDT) aka the Bradley Method

Aim to restore and maintain a satisfactory ecosystem whereby natural regeneration can occur. It appears slow because it is labour demanding but is extraordinarily effective in the longer term and will achieve more than herbicides.

Principles Work from least infested area to the most infested. Minimise soil disturbance. Allow native plant regeneration to dictate the rate of weed removal i.e. do not overclear, or allow the amount of time needed to consolidate the already weeded areas to determine the rate of primary weeding. Consolidation by reweeding initial flushes of weeds is crucial.

Suitable areas Where plants can colonise the site by seed or vegetable means. Areas sensitive to erosion. Areas likely to be overused.

Tools Trowel, pliers, secateurs, bow-saw, tree-loppers.

Methods of forest regeneration on degraded land

1. No replanting at all and allowing plants to re-seed windward.
2. Seeding the land by: Broadcasting, Hydro-mulching, Straw mulching, Brush transfer, Topsoil transfer, Kimseed camel pitter, pelletting.

3. Planting seedlings in various forms: Tubes, Semi-advanced, Advanced, Transplanting, Division of rhizomes and stolons.

Student activities

- *Students select an area of weed infested land; map the degree, types and causes of weed infestation, then decide on a method to regenerate a forest*
- *Place Zone V on their designs*
- *Visit a forest in good health and observe the ecological principles at work.*

UNIT 22
World climate biozones

Different climate regimes give rise to different indigenous plants, animals, soils and farm practices. Despite this, there has been mass transfer of farm technologies from cool temperature areas into tropical and desert areas with subsequent devastating land degradation.

This unit explains how broad climatic zones have evolved unique vegetation and soils. These require specialised strategies to be productive. Inappropriate strategies can be very destructive. Some results of misapplications of technology are:

- Irrigation salinity in arid areas
- Massive dam schemes in tropical regions
- Tree removal in tropical areas
- Broadscale chemical monocropping

Understanding the characteristics of broad bioregions and the special strategies required for them to be productive helps prevent the mistakes of the past. This is useful to permaculturists who want to work in other regions. It helps students to avoid errors of inappropriate water storage and use, cultivation and cropping techniques.

Over many centuries every climate and its ecosystems evolved over many centuries with people and animals and they have fed and clothed them. Each biozone developed appropriate agriculture. People lived sustainably in wetlands, deserts, steep hillsides, islands, torrid and very cold regions. It is only recently that there has been large scale destruction of ecosystems. Often this is because people have abandoned old methods and techniques and taken up new and inappropriate ones. There are also other reasons for ecosystem collapse but this is a main one.

If we learn and select the best of traditional practices perhaps we can adapt it to today's conditions and assist with Earth's survival with less shock.

Learning objectives

At the end of this unit, students will:

- Identify unique differences in each zone
- Describe how water, soils and nutrient banks are related in biozones
- Design water harvesting systems for each
- Explain what farming methods will be most productive in the areas where they may work and live

Teaching tools

- World globe or map to show the biozones
- Sketches of:
- Temperate climate glacially formed landscape
- Wet climate of rain formed landscape with rounded slopes
- Arid wind formed landscapes with mesas, buttes, and inland deltas
- Drawings showing cultivation techniques and traditional strategies

Terms

Although some of these terms have been used earlier it is useful to revise them.

Aquifer A body of saturated rock through which water moves easily.

Biomass The total amount of living material per surface area.

Caleche A pan which develops on tropical and arid soils after their forest cover is destroyed. They also occur in arid areas.

Chinampas Cultivation system of wetlands where fingers of land jut out into water areas – they were very productive in Ancient Meso-America and are in southern Viet Nam today.

Evapo-transpiration The combined loss of moisture from a plant from evaporation and transpiration.

Grey water Water that has been used once for any reason.

High dams High dams are used for irrigation and house water. Fed by gravity.

Low dams Low dams are used for aquaculture and irrigation.

Keyline The contour line leading out from a keypoint.

Swale A ditch or saucer dug along a contour line to allow water to infiltrate soils.

Ethics

Implement special strategies and techniques according to each broad climate zone and its requirement derived from its evolution.

Principles

- Research climate, soils and traditional water strategies of all new regions before introducing new technologies
- Remember that the interrelationships will be greater than you foresee so start small and go slowly
- Think in terms of local materials and species

- *Ask for examples of inappropriate farming practices*
- *Ask where the poles and the Equator are using a globe, map or draw a large circle for Earth. With colours, roughly sketch in the three zones mentioning patterns of climate e.g. East and west coasts of continents, and, northern and southern hemispheres. Link to the unit on Climate*
- *Divide the class into three groups and ask them each to take one of the zones. They will discuss it according to the headings on the left. Give about 10 minutes. Then fill in the gaps*

---

or

Take a different coloured paper for each biozone i.e. Tropical, Temperate and Arid and then write the major headings under Topics e.g. Soils, plants etc. as elements on each sheet. You will have three by eight items for each coloured page. Mix all the headings and then give them to each of the three class groups.

What design strategies do students suggest for overcoming limitations? (5-10 minute open discussion)

---

1. Humid-wet tropical or temperate landscapes

These landscapes are rounded because they were formed by water. This shape determines how soil, water, frost, and fire react, and, it controls design strategies.

Humid climates can rapidly deteriorate to semi-arid if forests are removed.

Show diagram of traditional Humid Landscapes.

Upper slopes

Forest left forever where soils are unstable at more than 18°.

Keypoints

- Forests create thermal zones and manage altitudinal changes
- These slopes are collection points for water (dams) and power
- Critical place for water control for lower slopes
- Diversion dams at keypoint
- Irrigation canals from keypoints along ridges
- Cultivation below keypoints
- Link keypoint-to-keypoint along keyline
- Housing is suitable here if forest is above so there is clean water in, from above and greywater out, to below

Topics	Tropical	Temperature	Arid
Soils	20-25% nutrients Little humus Leached fast No mulch develop	90-95% nutrients High organic matter Slower leaching Natural mulches	Plentiful nutrients Little OM Often salt No mulches
Plants	75-80% nutrients Biomass critical Stacking Use tree crops	5-10% nutrients Humus vital Deep-rooted crops Deciduous for cycling	Water release Mulch needed Deep-rooted crop Ephemerals
Landform	Humid rounded Glacial angular	Humid rounded or	Wind eroded
Water	Aquacultures	Ground storage	Underground
Biomass	Continuous growth Rain triggers germination and flowering	Harvest cycle Daylength and temperature regulate germination and flowering	Periodic Plants escape or tolerate dryness
Cultivation	Disastrous Caleche develops No tillage	Grain cropping	Spot strategy Opportunistic
Strategies	N-fixing ground-covers Nutrient cycling Stacking	Use heat and light Add to mulch Glasshouses	Drip irrigation Trees Shade
Limits	Soil poverty Heat Introduced crops	Temperature and light Fire, frost, water Chemicals	Dry periods Temp. extremes Overwatering

Flatlands

- Various irrigation layouts and techniques
- Use heavy mulches on flats
- Swales intercept run-off
- Use spiral earthbank designs
- Build check dams
- Protect creeks and rivers

Lower slopes

- Mixed crop cultivation areas
- Ensure fire control
- Build terraces and mini-terraces if slopes are steep

Illustrate the following concepts clearly and link to keyline design.

Treatment of individual slopes

- Steep and stoney; use net-and-pan structures
- Steep and grassy; place houses, swales, windbreaks, banks
- Very steep; build classical terraces

2. Arid landscapes

These landscapes are very complex. Desertification happens not only where land is turning to desert, but also anywhere the land produces less.

Obviously, the big problem is lack of water and, many soils are alkaline having a pH of 8.0-8.5. There are three considerations:-

- Settlements in Arid Areas
- Land shaping
- Planting and grazing

Settlements

In the Southern Hemisphere, settlements should be placed, if possible, on the lower southern or eastern slopes of hills. These are cooler, sheltered from hot northerly winds, and there is less evaporation. Here there is also the ability to capture run-off water, and soil carried down by water, wind or gravity and the plain is likely to be richer and finer. There may also be an aquifer or water lens. Here many strategies can be imposed to harvest water.

Land shaping is altering the land in ways to capture more water and make the land more productive. The following techniques can be used on marginal lands as well. Think in terms of small water systems, some of which may eventually dry up.

Swales used in arid landscapes are 3-4m (9ft10in-13ft) wide and up to one metre deep and from 100 to 150m apart. The length and depth of swales are calculated on the probable rainfall. If rainfall is 25.4cm (10in) per year and 1/3 runs off over 150m then 127cm (50in) is collected.

Use meanders to create ox-bow lakes, or billabongs. Swales and channels can be built to hold water, and to catch silt.

Hills trap water at the top of buttes and mesas or channel it along hillsides. Similarly, rocks can be used to achieve the same thing. At the bottom of hills, floodwater can be trapped in micro-catchments. Rain on sandhills or mesas gather a water lens underneath them. Plant on the edges.

Planting and grazing techniques

There are many of these, limited only by imagination and creativity. Some of the more common are given here. The basic principle is to concentrate on edibles. Arid areas can grow peaches, apricots, grapes, melons, Arabian coffee, and small quantities of grains such as millet, wheat and all the indigenous edibles for each country. In Australia, foods such as bush orange, native passionfruit, quandong, will all grow in arid areas. Plant two Acacias for every fruit tree. *Acacia victoriae* produces copious amounts of high protein seed which can be ground for food and is also a heavy leaf crop.

Always plant along **swales**.

Plant in plastic bags filled with compost, manure and sand which is very wet, and the bags have good drainage holes. Plant these deeply with a sand mulch on top.

Pit the land surface with two shovels, a disc plough, or the Kimseed Camel Pitter (a light weight machine pulled behind a ute or car for pitting and planting seed), pit the surface in a spiral pattern. Wind can blow from any direction – pits collect light fine nutrient rich clays and any organic matter; seeds are slightly protected when they germinate.

Use mulches and even river sand or stones will do. Water below the surface where possible with drip irrigation – only replacing water as the plant loses it.

Other landscapes

Other examples of environments which need special designs for productive outcomes are: Coasts, High Islands, Low Islands.

Swidden Slash and burn on an eleven year cycle. Quite sustainable but breaking down due to logging and reducing the necessary cycle time.

Wetlands Estuaries Cork-Pork Forest: Portugal

Chinampas are highly productive ditch-and-bank system in which banks extend out into water to maximise the productive edge. Examples exist in Mexico, Thailand, Viet Nam – primarily hot wet landscapes and deltas.

Silt round swamps and marshes. Land/water nutrient exchange give a harmonic effect. Use wild rice as main crop (*Zizania aquatica*). Occasionally drain areas and use mud for fertiliser onto the banks. Grow mussels and fish. Ducks are the main livestock for nutrient cycling and return of potash. Most fish are edge feeders. Azolla, a land mulch also grows in water. Trellis crops over the water economise on space.

From Wikipedia, the free encyclopedia:

Often referred to as 'floating gardens,' chinampas were a type of artificial islands that measured roughly 30×2.5 m (98 × 8.2ft), although sometimes longer. Originally they were used by the ancient Aztec Indians and were created by staking out the shallow lake bed and then fencing in the rectangle with wattle. The fenced-off area was then layered with mud, lake sediment, and decaying vegetation, eventually bringing it above the level of the lake. Often trees such as ¯ahuex¯otl (Salix bonplandiana) and ¯ahu¯ehu¯etl (Taxodium mucronatum) were planted at the corners to secure the chinampa. Chinampas were separated by channels wide enough for a canoe to pass. These 'islands' had very high crop yields with up to seven crops a year.

Student activities

- *Redesign an area that they know e.g. Coast or humid hillside with the naturally occurring plants and animals*
- *Describe the soils and the nutrient banks*
- *Design a suitable water harvesting system for the area*
- *Suggest farming strategies and yields*

SECTION 5

Increasing resilience and productivity

UNIT 23
Site analysis

This site analysis is the first step in the group design that students will present at the end of their course.

They are now asked to change perspective and look at land that will contain all five zones. Some students find it difficult to think in larger dimensions so they need more time to think. Students may need to spend additional time out of course teaching hours.

Review earlier design methods and the use of the topographic maps.

In this task students must do the best possible permaculture design. Each group will present their design as a team effort on the last day of the course with every person in the team presenting a separate aspect e.g. Site analysis, or Zone planning or water audit. The client, who has given the brief for the permaculture design, will be present for the presentation of the designs. I usually do this unit on site analysis about six days into the course. This gives students more time. We have finished the zones and other design elements such as aquaculture, and disaster resilience can be added in over the last six days.

The group site analysis and design is the major assessment assignment for the Permaculture Design Course certificate. You will assess students individually, however you have been aware of their progress from their home designs, and in class discussion and participation. Anyone falling behind has been monitored and assisted to reach class standard.

They have completed a step-by-step site analysis for their own land beginning with sector analysis and then water, soils, vegetation, climate and microclimates on that site. The site analysis is most important because it brings together all the information they need to do the final design. If students do a poor site analysis then their design solutions may be inaccurate.

When students correctly observe and define the problems and potential of the land, their design is likely to be resilient and productive.

Learning objectives

Students will:

- Know what to take on a site visit
- Connect information such as map reading and climate
- Ask relevant questions of the owner and develop a brief
- Make useful practical observations on-site
- Know where to collect off-site information
- Start drawing up plans for site analysis and theme design.

This unit will enable students to:

- Consolidate the skills they have been developing by applying concepts to a piece of land which they own or know well
- Use these skills to design an unknown site of different scale, preferably in a rural area but could also be a suburban neighbourhood or city block

Teaching tools

A variety of maps and in particular, a good site analysis.

Map and revise the sector analysis and microclimate analysis e.g. Soil/water/vegetation/animals/wildlife/structures.

Terms

The Brief Is a statement of what the client expects of you, and what you will provide for the client. If the work is very expensive, or difficult, it may require a legal contract.

Cadastral maps Maps which show accurately the extent and measurement of every field and plot of land. It is a register of property to serve as a basis of land tax (rates).

Catchment Ridge and valley run-off caught in one drainage system.

Contour lines Imaginary lines on the ground which join places of equal height above sea-level.

Topographic maps Maps having detailed description and delineation of any locality. There are many types.

Ethics

Approach a site respectfully as a complex set of interacting elements.

Principles

- Thorough and accurate site analyses leads to clear enunciation of the land's problems and potential
- Permaculture design entails imposition of zone and sector planning as appropriate for each piece of land
- The best site analyses are carried out by teams

- *Ask students to put themselves in groups of five or six people. It is important that each group knows and understands the potential and special abilities each team member brings, e.g. One is good at drafting, another at broadscale ideas, another at plant selection and so on*

- *Recall the principles behind Sector and Zone design. Revise overall energy, soil and water audits. Students list what they need to take on their first site visit and for a client interview*

What to take on a site visit

Stationary*:* Pencils, pen, paper for sketching, rubber, ruler, sturdy clipboard or small computer. Camera.
Tools: Tape measure of at least 10 metres, preferably 30 metres. Jars for water samples, plastic bags for soil and plant samples, shovel, hand trowel to test soils.
Dress: Waterproof boots/shoes, hat, sunscreen, water.
Off-site information: Climate data, history of land, zoning, land tenure, future local planning, neighbouring activities.
Maps: the best are topographic and aerial.
Weather information: Meteorological Bureau – rainfall, sunrise and sunset, wind speed and direction, maximum and minimum temperatures, evaporation, frost-free period etc.
Disaster information: Department of Water Resources for one in 100 year flood, season water courses, marshes and swamps. Fire frequency.
History of the land: If known.
Finally: List what the client supplies and will own when the job is finished. Have students compose the Brief with the client. If the client is not a permaculturist you may have talk to the client first about the scope of permaculture and what they can expect from the design.

Discuss information for students to gather on site. The following list can be used as a worksheet.

On-site information
Name and Address
Phone/email
Area of land
Site boundaries
The Brief
Client skills, knowledge, practices. Family and involvement.
Idea of financial status.
Time expectations; Hopes and expectations.

Climate	Wind Seasons Directions Strength	Rainfall Seasons Distribution Evaporation Frost/snow Humidity Hail	Radiation Direct Indirect Shade Captured Temperatures (averages and extremes)
Topography	Altitude Geology	Slopes Contours Drainage Keylines Gradients	Aspects Solar Shade
Water	Springs Lakes Ponds Bores Dams Wetlands	Rivers Creeks Flooding Water table Quality	Flow Seasonality Water capture potential from buildings etc.
Soils	pH Depth Types	Chemicals used History	Fertility assessment e.g. Organic matter
Ecosystems	Communities Edges Stability Niches Habitats	Health Weeds Wildlife corridors	Preservation orders
Fauna/flora	Pests Paths/routes Numbers Damage	Weeds Persistence Invasion Degree	Action taken
Natural Resources	Vegetables Fruit	Timber Animals	Other
Risk/Disaster profile	Fire	Floods	Economic etc.
Access	Vehicle Direct Damage Parking Condition Gradients	Maintenance Usage On contour	Other
Services	Telephone Water Sewerage Septic	Postal Web	Grey water
Structures	Houses, sheds Special places	Animal housing needs	Condition Extensions
Off-site constraints	Views Privacy	Pollution Remoteness	History Future needs Recreation Education
Legal constraints	Height, rights of way Set back Zoning	Future controls/ plans	

Project presentation

1. Student site analysis will include:
   Sector analysis, vegetation, energy, soil and water audits. Microclimates, wildlife, pests, disaster/risk assessment.
   Photos, soil and plant samples, and models of the land are all acceptable.
2. Theme Plan is drawn up on basis of site analysis findings. It comprises:
   Zones 0 to V, water energy, nutrient, soil, waste, windbreaks, micro-climates, disaster-resilience, enterprise possibilities, pests and wildlife, whole site assessments are basic inclusions.

Student activities

- *Meet in groups and draw up a base plan, adding to it as more information becomes available.*
- *Finish with a full site analysis and a separate theme plan for the site. Models and photographs are useful. Students need time between the site visit and producing plans.*
- *Residential courses, or weekend part-time residentials enable students to get up at 5 am and work until midnight!*

UNIT 24
Design graphics and creative problem solving

This unit is to teach students to get their ideas on paper with some accuracy. The aim is not for them to become drafts people. There is no one method. However, it is important to stress that designs must be readable – i.e. not just a mess of complicated lines.

There will always be some people who simply cannot do plans (left- or right-brain approaches perhaps?), however they can still be effective consultants. They can know a lot about on land rehabilitation, by walking over land with the owners and showing them where, and how, elements should be placed, rather than transferring their ideas to paper. Very often this is all the landowner wants.

If plans are required for management, or to be sent to Council, then good plans must submitted.

Drawing up plans is simple. The really creative work is done by working out the design. Putting it on paper could be done by taking the ideas to a drafts person.

From now to the end of the course students will work together in groups to draw up plans for the brief.

Students have two topics in this unit:

- The first topic is design graphics about developing base plans and clarifying the brief for the larger piece of group work
- The second topic is about creative problem solving and is a discussion on creative thinking and approaches to problem solving

1. Design graphics

Learning objectives

In the unit on Map Reading students started to learn about scale and maps. This unit is largely practical. Students will experiment with different scales for large areas of land and choose the appropriate scale for their piece of work. It is important not to confuse good art work with good design. It is very easy to see brilliant art work as demonstrating design principles. Sometimes they do not go together.

Have students experiment with their strengths and weaknesses; to be brave enough to try new ideas such as models or collages. Not enough students use photos.

Computer designs with Sketch-up or other programs, are now appearing.

At the end of the session students will be able to:

- Compare the usefulness of different scales
- Enlarge a base map to a desirable scale
- Roughly place elements in their correct positions e.g. buildings, creeks
- Use several types of pens, rulers, t-squares, and keys

Teaching tools

- Copies of a variety of past designs which are clear and unclear
- Show good examples of a Site Analysis and then a final design
- Show examples of naif art. Vietnamese and Cambodians do wonderful, clear drawings of this type because art as we know it, was not taught in schools
- There are no special terms for design graphics
- Ensure all students have an opportunity to work on the design and repeat that each one must speak about the part that they took responsibility for

Principles

- Drawings should be easy to read
- Scale needs to be appropriate
- Construction drawings are usually not necessary
- Do the site analysis before final design

It is not the artwork which is important but the placement of elements and their connexions with other elements and the land.

Drawing strategies and techniques

This is a show and tell. Bring a box of drafting goodies.

Demonstrate some designs in different styles and sizes (discuss the best way to approach design).

Start with paper sizes from quarto to A1, talk about the different uses of each and include graph paper and tracing paper:

- Show a series of pens and discuss thickness, waterproofing, and erasability
- Show a series of pencils, from about 2H to 6B and how each is used. Discuss different types of erasers at the same time
- Demonstrate the use of different implements such as drawing boards, set squares, t-squares, and rulers
- Stress that printing must be even and that capitals are the most effective. Rule top and bottom lines
- Point out that computers are useful for print. Stencils exist for graphics such as tree silhouettes. Students can copy ideas from books, web and other plans
- A cross-section also known as a transverse section is effective to illustrate a point and free-hand designs also help with concepts
- Show plans with the conventional identifying Titles, Dates, etc. and where these are placed. Draw attention to thick pen used for boundaries and thinner for contours. Show the conventions for scale
- Explain what is required by a site analysis – dividing the area up into natural regions and listing the problems and potentials within each, so as to ensure that the correct design solutions are given
- Next describe a Master Plan or Theme Plan. Show what is expected
- Mention that construction drawings have great detail and are sometimes required. They give sufficient detail to be able to take the costings off it
- Some clients want a written statement of the visit as well as plans and other clients cannot read plans well. So Consultants must go over plans very carefully with clients

2. Creative problem solving

Creative problem solving is the mental process of creating a solution to a problem. It is a special form of problem solving in which the solution is independently created rather than learned with assistance. Creative problem solving always involves creativity.

To qualify as creative problem solving the solution must either have value, clearly solve the stated problem, or be appreciated by someone for whom the situation improves.

The situation prior to the solution does not need to be labelled as a problem. Alternate labels include a challenge, an opportunity, or a situation in which there is room for improvement.

If a created solution becomes widely used, the solution becomes an innovation and the word innovation also refers to the process of creating that innovation. A widespread and long-lived innovation typically becomes a new tradition. 'All innovations [begin] as creative solutions, but not all creative solutions become innovations.' Some innovations also qualify as inventions.

Inventing is a special kind of creative problem solving in which the created solution qualifies as an invention because it is a useful new object, substance, process, software, or other kind of marketable entity.'*

There are no problems, only solutions. Design is creative problem solving. This unit coordinates with Unit 23, Site Analysis, and will be useful for completing the final design. Students are encouraged to respond to new ideas; to challenge, to question and to reflect on their own experience. Permaculture is creative and innovative.

Most people think in the most boring ways about everyday matters. They tend to think only about the same things and when they do, it is what they heard or read, or what their parents said. This is called grooving which is what happens when a needle goes around and around playing a record.

So in doing permaculture designs, people tend to think only of the same sort of answers. Creative thinking has brought about exciting changes. Encourage students to have different ideas.

Fear is the main reason why some people don't think creatively. It is a major block to creativity.

- Fear of making mistakes
- Fear of being laughed at
- Fear of being wrong
- Fear of being alone
- Fear of criticism

Pride makes it important that we don't suffer these things.

Learning objectives

To encourage lateral thinking and to open up design options are the goals of this short unit. Students are encouraged to:

- Acquire objectivity
- Be wrong and consider it valuable
- Think widely and wildly
- Take chances

Teaching tools

- The two faces/vase drawing to see how although the lines are exactly the same and one person sees a face while another sees a vase
- Take a match box and ask how many uses there are for it? Encourage really different answers
- Explain the Brainstorm ... Do a Rausach Blot test

Terms

Creative solutions Lateral thinking.
Cognitive dissonance Mental indigestion or unease.
Constructive discontent Dissatisfaction for a better result.
Problem The way to a design solution.
Protective cognition Denial.

Ethics

What others know and contribute is equally important as your own. Many very creative ideas and works come out of team work e.g. Designers, film-makers. A good team will do better work than a single person because of the number of variables in permaculture – everyone is good at some part of it.

Principles

- Believe in and trust ourselves. This is not arrogance
- Recognise constructive discontent is feeling dissatisfied or uneasy about something and living with that unease until the answer is finally realised. It is more subconscious and instinctive and useful in helping extend conscious limits
- See the whole picture and recognise your habits of behaviour, thought and response. Stand aside from self
- Put yourself in someone else's shoes. Look at the situation from the opposite end of the spectrum – e.g. Developer, the real estate agent. Think there are no limits and see what happens
- Recognise your own denial as protective cognition. Let it go

Steps in creative design

From : Boosting Your Creativity Ability from www.mindtools.com/pages/article/creativity-quiz.htm

In his well-respected book, *Creativity*, Mihaly Csikszentmihalyi says that an effective, creative process usually consists of five steps. These are:

1. Preparation – becoming immersed in problems and issues that are interesting and that arouse curiosity
2. Incubation – allowing ideas to turn around in your mind without thinking about them consciously
3. Insight – experiencing the moment when the problem makes sense, and you understand the fundamental issue
4. Evaluation – taking time to make sure that the insight provides sufficient value to outweigh the various costs involved in implementation
5. Elaboration – creating a plan to implement the solution, and following through.

* *Lateral Thinking*, Edward De Bono; Penguin UK, 1974

Steps in creative design from Bill Mollison

Acceptance

- Accept that there is a problem and take responsibility for it. Accept that you can do the design and ask these questions:
- Do I have the time?
- Do I have the ability/expertise?
- It is a mistake to wait until you are absolutely sure you have a design/solution before beginning.
- With such certitude you enter the problem solving process with all decisions made and all problems solved.

Analysis

- Question everything
- Compare e.g. Discover relationships between things e.g. Sunshine, wind, insects, birds
- Dissect, breakdown, explore, integrate, find sequences and patterns

What others know and contribute is as important as one's own thoughts. Use tools like models, brainstorms, connections: stretch, expand. Leave it alone. Go away. Come back later.

Define

- Translate the real world into symbols. i.e. increase the precision of your thought and definition
- Progress from unclear to very clear statements of a problem e.g. Cold wind...cold winter wind ...cold July wind....Cold July wind for one week

Form ideas

- Brainstorm alone or with others because there are no buts in a brainstorm
- The most absurd ideas can trigger off something really practical
- Research ideas – library, other people, information
- Have a break

Sort and select ideas

- Redefine your idea. See how it fulfills the definition of the problem. Too much favouring of a pet idea?
- Evaluate for relative worth

Implement

- Sketch design on paper and evaluate it by checking with 'on the ground' reality
- Implement concept plan when satisfied with ideas and their relationships

Reflect

- Look back and reflect on the value of what has been done
- Detect the flaws and discoveries
- Detach at this time
- When working with a client, remember you are not the owner
- Don't be possessive of the design/project (Easier said than done)
- Don't overvalue your ideas
- Don't be afraid to use intuition
- Meet your starting objectives
- Don't be afraid of the bigness or smallness of something. A big design is made up of many small systems, small is only part of something much bigger than itself

Design is

- Finding the optimum in a particular set of circumstances
- The interface between people and environment

Creativity is just a matter of getting rid of something; some limiting factor.

Three Levels of Human Awareness

1. Attraction
2. Understanding – developing meaning
3. Manipulation – using the meaning

Three Levels for self control and creative design

1. Relaxation
2. Imagination – let it run riot
3. Concentration (centredness) – bring attention and effort to the situation, avoid distractions

Basic Dreams and Expectations of Modern People

1. To put energies into creativity and something constructive
2. To have an imaginative lifestyle – outrageous even
3. To have a complete, loving life (sex included) – not to be without close human relations
4. To be in control

Student activities

- *Try ideas from this unit while working on group and individual designs*
- *Simply do a design to try these ideas*
- *Surprise yourself*

UNIT 25
Incomes from acres

The permaculture principles of diversity and stability also apply to earning income, as do strategies of working to minimise maintenance and enhancing one enterprise with another. All land should pay for itself i.e. pay all rates and taxes and give some return for living on it.

Some people start immediately to establish farm/garden enterprises; others work to a gradual, rolling permaculture; while another group feels stuck with wheat, wool and cattle and don't know how to change.

A landscape design is incomplete without offering some ideas of enterprise development which will increase income or give a parallel one. This unit accompanies that on Site Analysis and later on Ethical Money and the concept of right livelihood.

Land can pay for itself by you using the skills that you have learned on that land, or, by developing products. So for example, you can build wind generators because you have built your own, or you save seed from your own land. Land should never be abused or degraded which means having its abundance, resilience and sustainability reduced. Good design enhances and protects enterprises.

With fragile world economic systems it is important to be realistic and put work into income generation. Although there may be economic collapse, it can also present opportunities to meet human and animal needs.

Learning objectives

By the end of this unit, students will be able to:

- Offer considerations for generating income
- Discuss the marketing and processing of goods
- Consider some alternative enterprises
- Explain two model situations

Terms

Income Earnings from the land which may be on-farm or off-farm but derive from living on that land, for example, teaching about organic farming from one's own experience

Teaching tools

- Demonstrate examples of appropriate technology, and people creatively earning money from their land
- Use posters, drawings, DVDs or web video clips or demonstrate the products themselves
- Involve students in a brainstorm for this topic. Students enjoy it and it generates a huge number of ideas

It also raises the class energy if they are tired

Brainstorm for income generation ideas. Ask two people to write on the board and appoint a time-keeper. At 'GO', everyone calls out an income generating idea. They have ten minutes. There are no rules except ideas which are obscene or offensive. At the end of the time see how many different ideas there are. Normally there are about one hundred. Now students take pens and link those which integrate. They eliminate the ideas which are utterly unrealistic.

Ethics

- Care for the land and develop enterprises which also care for people
- Reduce consumerism and meet needs
- Replenish renewable resources

Principles

- The land pays for itself
- Minimum maintenance by design of ecosystems
- Achieve long term sustainability
- Start small
- Diversify to no more than four enterprises.

Revise the concept of diversity and show its relevance in diversifying incomes so as to reduce economic risk.

Briefly revise Zones 0 and I for siting structures and home vegetable gardens to avoid damaging heat, cold, fire and to favour animals. Ensure that the buildings work with the others and with the land.

Considerations for income generating activities

Start small. Even part-time and build up. If you go big immediately you can be stuck with something you didn't want or won't work.

Diversify your sources of income.

Be realistic about your own capabilities – learn from the mistakes and successes of others.

Be ruthlessly honest with yourself about what you want to do.

Know when to seek expertise.

Work out **short and long term strategies**. For example, income for self (first), and children (medium term) and their children (long term) and then extended family e.g. Some precious timbers need a 200 to 400 year plan.

Look at long term goals as closely as short term. Search out local resources e.g. Timbermill for sawdust, fuel; chicken factory for manure. And together the manure and sawdust are excellent fertiliser when composted correctly.

Potential enterprises

Marketing produce

Select for maximum value for weight and bulk of produce e.g. Nuts, honey and dried fruit. These store well and can be strategically released onto the market. Grow grains for own use not for the market because of land degradation.

Follow up by probing questions about why it is preferable to grow grains for own use.

Perishable products can be either directly marketed from the farm gate or self-picked, or sold wholesale to local producer/farmer markets where the producer sell directly, or to huge city markets.

- Co-op growing subscriber networks with regular deliveries
- Rent-a-Tree, set up orchards and rental pays establishment and maintenance
- Mail order for fresh and dried herbs, worms and seeds
- Rent-a-duck for snail eating
- Rent-a-sheep for lawn mowing, animals are rented at $5.00 per week. Owner pays vet fees and shearing costs

Drying and storing fruits

- Best crops are root crops and dried fruits such as apricots and peaches
- A drying shed is easily built on stilts and 1m high with a solar angle of 31° for the roof (Sydney area). Racks are easily made from old fly screen doors
- Build a solar chimney (a black pipe) to pull warm air through over the fruit

Herbs and flowers

Herbs and flowers are dried in the dark for herbs to retain healing properties and essential oils. (Fruits can be dried in the sun.) Bundle herbs and flowers and hang on hooks upside down in a shed.

Australia still imports 95% of its herbs. Manor Herbs (world growers) are looking to Australia to supply clean stock.

- When growing herbs, the biggest problem is weed competition. Weeder geese will manage by eating grasses
- Market research is important to select appropriate herbs
- Analyse soils and product needs and add minerals and nutrients if required
- As well as drying herbs, produce tincture, ointment, cosmetics and other products

Seed for sale

Seeds are saved for food and sale. Wrap seed heads in brown paper bags to catch seeds. Paper breathes well and absorbs moisture (prevents fungal diseases). Soya beans, dandelion roots and some wattle seeds are roasted and ground for a beverage.

Distilled and pressed oils

Cold pressing of oil is fairly simple, full details of the technology is in many libraries and online. Distilling is also simple. Put harvested crops of herbs, e.g. Lemongrass, lavender, in a big bin of water, bring to boil and have two outlets, one for steam and one for oil. Tea-tree and eucalypts have long been harvested for oil.

Waste recycling

Non-marketable apples and other fruit can be used for vinegar, pickles, bottling and juices.

Tree prunings are good for smoking foods such as nuts, meat and cheese. Hickory smoked goods fetch premium prices.

After fires, dead and dangerous trees were milled, or mulched using a portable mill. Some valuable timbers were recovered.

Specialist work

- Contract tree planting, and designs
- Insect breeding e.g. Ladybirds for biological pest control of aphids
- Earth moving for those who are excellent at it.
- Seasonal agistment of animals, especially during summer growth
- Animals for special purposes e.g. sheep for carpet wool, exotic poultry and eggs, milk sheep for cheeses (fetta and rocquefort) and even cannibal snails
- Alternative energy technology for solar panels, ovens, barbeques and wind power and pedal power (Peter Peddles, NSW) for wheat grinders, washing machines, blenders etc
- Alternative Real Estate to include camping and fishing on farms, farm holidays, courses and education, firewood, or even picnics. Hideouts for artists and musicians

- Computer and software systems for alternative businesses, caring for mailing and membership lists, stickers and updates
- Information sales – publication, teaching courses, selling designs, etc
- Selling skills such as mudbrick and making sheet mulch gardens
- Water specialists – grey water cleansing, swimming pools, solar pumps, water plants, and water reticulation
- Specialist crops such as bush foods, and plants. Also permaculture nursery with nut and forest trees for mulches, fuel, pulp etc
- Special film locations – horse carriages, old stock e.g. Draught horses, wildflowers, fishing, refuges
- Hunting feral animals – Wild pigs in NSW fetch $100 on the German luxury food market as chemical free meat

Remember

Try to fill a market gap. Produce a quality product. Do your market research. Specialise.

Student activities

- *Work out a plan for getting some dollars from rogue land. Revise the Considerations for Enterprises' list and come up with priorities*
- *In their final design suggest how the marketing can be carried out for the recommended enterprises.*

UNIT 26
Design for mitigating disaster

Daily reports of disasters all over the world are reflecting the reality of global climate change. Disasters come in all shapes and sizes and occur in all places. Many are unpredictable.

Designing by Zones in permaculture decreases work, lowers energy inputs, increases yields, and stability of production. All these can be lost however, if the risk of disaster is not considered and resilience factored in. So it is important that students add disaster analysis and protection to their designs. This will strengthen the stability of their whole system.

Disasters can be both man-made and natural. Most are followed by secondary disasters such as epidemics or famine.

Disasters:

- Are situations which affect communities
- Result in serious effects on those communities
- Demand a total community response and action

Learning objectives

On completing this unit, students will be able to:

- List several types of catastrophe
- Make a disaster/risk profile
- Describe the causes, frequency and duration of several disasters
- Design landscapes and site buildings which will evade or survive catastrophes

Teaching tools

- Revise the coastal windbreak belts and root vegetables for cyclone areas
- Review Zones I to V and discuss where to place emergency small gardens
- Drawings of a 'fire-avoidance-resistant' garden. Decide the most likely disaster in this bioregion

Terms

Controllability Exercise restraint or direction on the free action of something.

Predictability To foretell or announce beforehand from statistical analysis.

Ethics

Save lives first and property last.

Principles

- Carry out a risk analysis of most likely occurrence
- Do a disaster analysis of the two most probable disasters
- Accept the precautionary principle – if it's likely to happen it probably will

Types of common disasters

Ask class for a list of possible disasters. Follow this with secondary disasters. Split the list into natural and man-made. The line between them is not always clear but working it out is a valuable exercise.

- Fire
- Famine
- Earthquake
- Landslide
- Epidemic
- Nuclear accident
- Chemical spills
- Disaster analysis
- Flood
- Cyclone
- Drought
- Tsunami
- Climate change
- War

Ask students for a list of questions they must ask before starting their design. Elicit the following:

- Cause of the disaster – natural or man-made
- Frequency – how often it occurs
- Duration – short or long time
- Speed of onset – warning period
- Scope of impact – concentrated or large areas
- Destructive potential – varies enormously
- Predictability – follows a pattern perhaps
- Controllability – are people helpless

Now have students suggest two of the most likely disasters for their bioregion and the fill in the table above. Work in two's.

Design for disaster avoidance resistance or endurance

Resistance Start with structures and apply the permaculture principle to reduce risk.

Endurance Store small spare supplies of seed, store plants and water away from the likely centre of the disaster. A cave, an underground room, a small mud house, or store on an island in a dam, are all good places to keep spares.

Avoidance Ensure there are escape routes such as creeks, fire trails, and green belts. Have a small emergency garden a long way from the disaster centre. Perhaps only potatoes, onions and hardy food shrubs.

Disasters type and specific design

Floods

These can be anticipated from weather statistics. Recently there has been heavy, unusual flooding and also extreme drought. This is part of global warming. Design and site all new structures higher than the 1:100 year flood contour.

- Have an emergency garden out of flood reach.
- Plant trees and shrubs densely by river banks to reduce the energy of the flood.
- Trees and shrubs planted through catchment areas reduce silting and hence flooding of rivers.
- If flooding occurs, do not enter floodwater on foot; use car, boats or wait to be evacuated. Climb to the roof.
- Use vehicles judiciously.
- Do not drink floodwater as it is often contaminated by sewerage. Carry bottles of boiled water. Do not panic

Cyclones – often accompanied by flooding

Houses need to be built with cyclone bolts and as close to the ground as possible, even underground. Or use a 45° roof angle, cut stud into brace and have it high pitched.

- Windbreak trees must be flexible – classically they are palms or bamboos which absorb a lot of the wind force. Small-leafed and multi-stemmed shrubs with good root systems are priority planting.
- Remain in shelter during and after the passing of the 'eye'. Have a 'famine' garden in a very sheltered area. Every cyclone is dangerous and must be treated as a real threat.

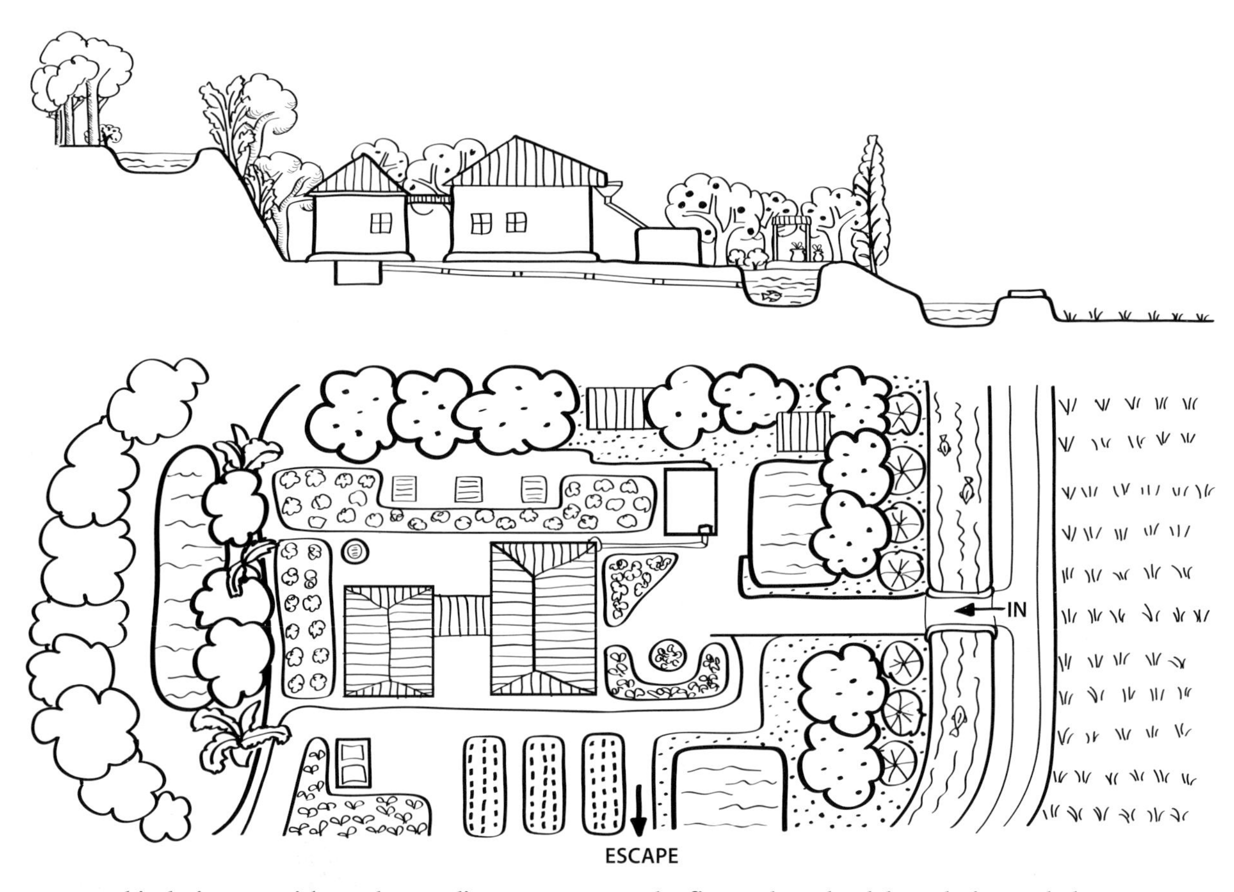

This design can withstand many disasters e.g. Drought, fire, cyclone, land degredation and plague

Droughts – often accompanied by famine

These are a normal part of many climates and there are peoples e.g. the Bishnoi in India, who live in close to perpetual drought conditions and consider this normal and their environment quite benign.

- Control stocks and consumption.
- Never to lose supplies of seed or animals through greed or carelessness.
- Water must be kept clean and not fouled.

Where drought is part of the normal weather cycle then pastures and feed can be saved. Many effective drought-proofing strategies and techniques exist so people and ecosystems can survive quite well through long periods without rain.

Example of techniques are in the *Permaculture Drylands Design Course Handbook.*

Nuclear Accident/Radiation – can be accompanied by illness

The biggest risk to most people is probably small accidents from leaking equipment with nuclear technology e.g. X-rays, irradiated food, planes, photographic industry etc.

Low level electro-magnetic radiation was researched by the NSW Government (1990). Results of the enquiry including compensation were out in February 1991. Limit consumption of electricity and 'consumer durables' as this assists in reducing health risk.

- After a nuclear accident, all water will be radioactive.
- Some water cleansing plants may be effective in reducing radioactivity (see Unit 9, Water).

If there were an accident at Lucas Heights, a south-easterly wind would carry dangerous substances over Sydney. Auroville, India, is in danger from the nuclear plant 100km up the coast and contaminated water could be carried by south-flowing currents.

Glasshouses After Chernobyl, fortunate people who had glasshouses had relatively uncontaminated food.

Miso Unpasteurised (raw) miso has enzymes to help the body eliminate long lived radioactivity. It was used with some success after the bombing of Hiroshima and Nagasaki.

Landslides, land degradation/famine

These have been fairly extensively covered throughout this Design Course. There are many known and tried techniques to repair land. In particular the application of Sector and Zone planning will alleviate problems and reduce risk.

Epidemic

In an epidemic the risk of contamination is through food and water. Isolation is the primary way of breaking epidemics, and home garden and rainwater tanks reduce disease transmission greatly. In effect, being independent of outside purchases is a risk reducing strategy. Epidemics run for predictable times then subside. The amount of food and water for this period is easily calculated and provided for.

Tsunami/earthquakes

In general these are less predictable disasters. There is nothing really to be done for a wave caused by earthquakes, volcanic eruptions and underwater explosions. There is usually a series of waves and any wave can be the biggest.

Earthquakes are having a greater effect on humans because of our increasing population. For example, a major earthquake took place in Northern Territory a few years ago which was recorded as quite high on the Richter scale. A lesser earthquake at Newcastle recently had a bigger impact on people and property because of the higher population density.

There is other evidence that earthquakes are occurring more frequently and are related to French underground nuclear tests in the Pacific. Earthquakes are also becoming more frequent in mining areas where geologists are now suggesting that rocks (coal, iron ore, etc.) are like buffers to movement on the surface of deep internal disruptions and that removing minerals in large quantities takes away the 'cushion' effect (ABC Science Show late 1990 or early 1991).

In the Earthquake Belt, large buildings are constructed according to strict regulations and people generally know to shelter in doorways if there is a quake.

Fire

Fire is a major threat in Australia. Design a small fire safe core area. Surround this with an intermediate buffer zone with fire safety features. Regenerating areas then surround these.

Factors in Fire Risk

Fuel	Doubling floor fuel produces quadrupling of fire intensity. Pine needles burn faster than thicker matter.
Mulches and bark	Dry mulches of annual grass, cereal crops, and pasture burn very fast. Fibrous barks burn more than smooth bark. Wet mulch in Zone I helps prevent fire.
Dry fuel and winds	Dry fuel and winds increase the risk of fire.
Topography	Fire moves faster uphill. Westerly aspects are often hotter.

Design in fire control

Use classic permaculture zone design diagram for design for fire prevention. Encourage discussion.

- Zone I garden – Around the house, damp mulches, green mulches, irrigated and green plants
- Water – Stored in irrigation/aquaculture ponds and tanks about the house. Plug and fill gutters, basins and baths. Have hoses inside
- Roads and paths – Keep clear. Lead away from the direction from which fires come
- Orchards – Are excellent fire breaks
- Plants – Fire retardant plants burn poorly, e.g. Coprosma, wattles, agapanthus, willows, carob, mulberries, figs, deciduous fruit
- Fire resistant plants – eucalypts which regenerate
- Fire shelter – A place people can escape to if the house burns – build of mud, rock, or inside of hill, and whitewashed
- Stored in irrigation/aquaculture ponds and tanks about the house. Plug and fill gutters, basins and baths. Have hoses inside
- Animal yards – Doors open to the cool aspect away from fire direction
- Radiant heat barriers – Stone walls, mud walls, earthbanks, concrete, bricks, thick low hedges, white walls, and fly screens on windows
- Firebreaks – Not cleared roads, but dense plantings of fire retardant and resistant species. In the Victorian fires these proved remarkably effective

Student activities

- *Estimate disaster risk to their own dwelling and to design protection*
- *How they would consider changing the plan if there were:*

- *a large reduction in the ozone layer?*
- *a change in climate causing cyclones or strong winds?*
- *a threat of radiation cloud?*
- *Ensure that in your own home plan and in the group plan you have considered the disaster most likely in your bioregion or micro-climate*

UNIT 27 Integrated pest management

Insects and weeds have a right to be there, a role to fill. Every species has the Right to Life and no species should ever be globally or locally exterminated. They are a natural part of ecosystems. If their numbers become unbalanced, it indicates that other conditions are out of balance. When soil and plants are right, the problem will balance out. This process will probably take three years from the establishment of a garden, or switching over to chemical-free techniques.

Pests are a sign that the environment is out of balance. The best pest management brings the cultivated ecosystem into balance again. It means farmers and gardeners must understand why the system is out of balance, and how balance is to be restored, and that means studying pest lifecycles and their predators. Integrated Pest Management is also called Intelligent Pest Management because it includes many variables and requires excellent observation and deduction.

Conventional pest management is based on prescription. All chemical commercial preparations are non-selective while most insects are NOT harmful. Australia has about 108,000 species, few of which affect our gardens, they are an important part of the food chain. They are pollinators, herbivores, carnivores and decomposers. Fewer than 1% of the resident insects in our gardens can be considered pests.

Learning objectives

There are many good books and a great deal of information in magazines. Encourage students to collect whatever is relevant data e.g. Pests of livestock or orchards, and observe more and more closely.

By the end of this unit, students will:

- Observe a pest, a species out of balance with its environment, and the conditions that cause this
- Describe the role of predators and parasites in pest management
- List the methods of control
- Explain the role that an insect's life cycle plays in controlling it

Terms

Parasite An animal that lives in or on another

organism (its host) from which it obtains its food.

Pest An organism, usually an animal which is irritating or troublesome to some degree – a species which is out of balance with its environment.

Predators Natural enemies of pests – any animal that eats other insects.

Nematode Small organism which is threadlike and usually lives in the soil and infests roots of such plants as tomatoes.

Pheromone A smell exuded by male or female animals to attract the other sex.

Ethic

Every species has a right to life and does not have to be useful or beneficial to humans.

Principles

- Pests are indicators that there is something wrong
- Pests are part of the diversity of life and require deterring not eliminating
- First, always aim to control insect pests by other insects or birds
- Aim to keep pests at a level where minimal or acceptable amounts of damage occur

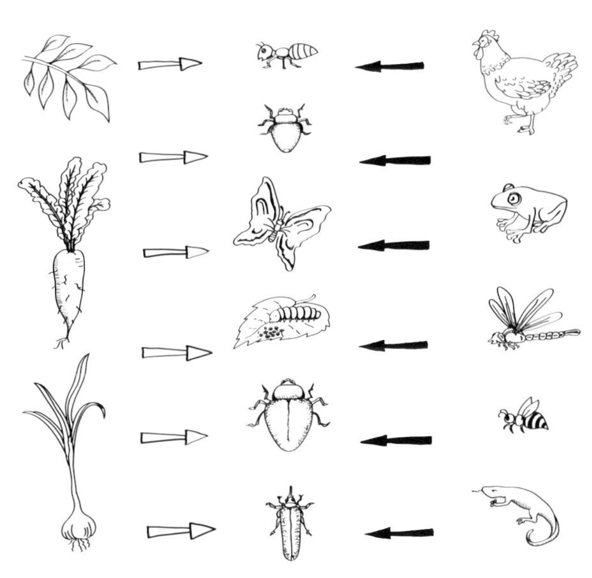

Plants, Pests and Predators

Causes of pests

Ask students to give examples.

- Monocultures
- New insect species e.g. Snails, fruitfly
- Predator destruction e.g. Where non-target pesticides are used
- Catastrophe e.g. Fire, clearing, flood
- Plant breeding e.g. New varieties attractive to pests

Methods of pest management

An effective pest management strategy requires designing systems which mimic natural ecosystems. And this diversity and structure creates resilience to many external or internal pressures which favour pests.

The diversity is reflected in such structures as: Water in several ponds for frogs, bees and fish; living fences of fruit, flowering species and homes, e.g. old logs for lizards and snakes; ducks in orchards; and mix herbs and vegetables.

Ask students for strategies to manage rather than to eradicate pests.

Ecosystem features

Encourage predators and parasites

Consciously design to encourage predators and parasites by increasing their habitats. Plant a variety of plants including ground covers, fruit trees, vines, perennials. Some of these will attract and shelter pest predators. Some plants will hide pest susceptible varieties.

Create habitats

Predator apartments

Useful animals

Birds

- Design habitats to provide safety from cats and large birds
- Plant many thorny shrubs to provide protection e.g. Pyracantha, Grevillea
- Offer several types of water e.g. Baths, ponds and sprays, and sacrificial fruiting species (sour, non- grafted fruit trees) as water for birds
- Use stakes of different sizes for birds to perch upon
- Birds eat moths, weevils, bees, lerps, aphids and flies. Small birds feed on a large percentage of insects
- Bantams and ducks feed on snails, some slugs, insects, cutworm
- Guinea fowl feed on grasshoppers, insects
- Let chickens in under the house to eat white ants
- Fish, toads and frogs need cool, moist places and eat pear slug, caterpillars and mosquitoes
- Lizards and centipedes need rocks and pieces of wood as places to live

Insects

A good garden is busy and noisy with insects working in it e.g. Bees, wasps, drones – all singing. Bees are possibly the most useful insect in a garden and are killed by any and every broad-spectrum spray. Lacewings are one of the most effective insect predators. As larvae they feed on aphids, psyllids, mealy bugs and moth eggs. Adults are small with gossamer wings and feed on nectar, pollen, aphids and mealy bugs. They are easily recognised by their eggs sitting on separate stalks.

All spiders do a good job of cleaning up pests.

Ladybugs, both adults and larvae feed on aphids, mealy bugs, scale and other small insects, (not the 28-spot).

All wasps are insect predators, or parasitoids. Parasitoids are even more effective than predators. Changes after being parasitised are often recognisable e.g. Whitefly eggs go from cream to black when parasitised.

Robber flies and hoverflies are useful predators. The hoverfly larvae are voracious and feed on aphids, mealybugs and mites.

Dragonflies feed on flies and mosquitos. Praying Mantis are excellent predators. Ground beetles feed on slugs, insect eggs and larvae.

Plant selection

Plant species to grow amongst vegetables which are known to encourage predators of known pests and/or repel pests. The plant families Asteraceae containing cosmos, chrysanthemums, aster, chamomile, artemesia and marigolds, and Umbelliferaceae, yarrow, anise, dill, angelica, fennel, parsley and Queen Anne's lace, provide food and shelter for parasites. Neem tree, see reference, is showing promise.

Mechanical techniques

These include baits, traps, barriers and lures to reduce pest numbers or interfere with their activities.

Barriers are sawdust, sharp sand, soot, cinders, ash around special plants or beds are barriers to pests such as snails and slugs as they are abrasive and/or dehydrating. Use tar compounds on pruning cuts.

Bands are wrapped around tree trunks to deter crawling insects and those pests which overwinter in the soil. Grease or resins can be used inside the bands.

Band lime to stop caterpillars around trees and seedlings.

Collars deter cutworm and can be made from

cardboard or beer cans with both ends cut out and pushed 5cm into the soil.

Traps include upturned citrus shells, sticky yellow boards, half bottles of stale beer. White attracts thrips.

Lures and Baits: Milk is attractive to slugs and snails. Yeasts, sugars and proteins are baits for fruit fly.

Put vinegar and sugar solution in traps made from plastic bottles with tops cut off and inverted. Ten to a tree will kill fruit fly. For this to be successful the neighbours may have to do it too.

Pheromones – insect hormones e.g. Dakpot.

Hand picking grubs or eggs daily is very effective and pleasurable.

Cultural controls

These keep pest outbreaks to a minimum and include:

Cover crops Crops planted as a sacrificial crop to remove pests from the soil before planting a susceptible crop. Mustard and rape remove wireworm from the soil.

Sanitation Removing and destroying diseased fruit and leaves can prevent many diseases. e.g. Black spot and fruitfly. Birds on the ground such as hens, quail, and ducks can effectively break these cycles. Keep rotted fruit with worms or fruitfly out of the compost heap. Seal fruit in a large, strong, black plastic bag and leave in the sun to cook. Then add to the compost heap.

Crop rotation Rotate crops to stop the build-up of pests/diseases in the soil. Generally, don't follow members of the same family in the same bed. Follow e.g. Don't plant tomatoes with potatoes.

Avoid peak season e.g. Plant early or late to avoid fruitfly in say, peaches, tomatoes.

Pest resistant plants Select local strong species. See reference, To Deter Birds, new research on breeding chemicals in to fruit which are unpalatable to birds but harmless, or not noticed, by humans. (Very suspect).

Interplant aromatic herbs and companion plants with vegetables. e.g. Tansy, pennyroyal, rue, mint, wormwood, rosemary, sage, lavender, basil, peppermint, southernwood, bay.

Manage soil by maintaining fertility. High organic matter reduces nematodes in soil. Proper irrigation and drainage (timing and type). Weed control without damaging roots.

Rampancy Use pumpkins to climb over potato vine or chokoes over lantana, etc.

Fish enjoy insects attracted to light

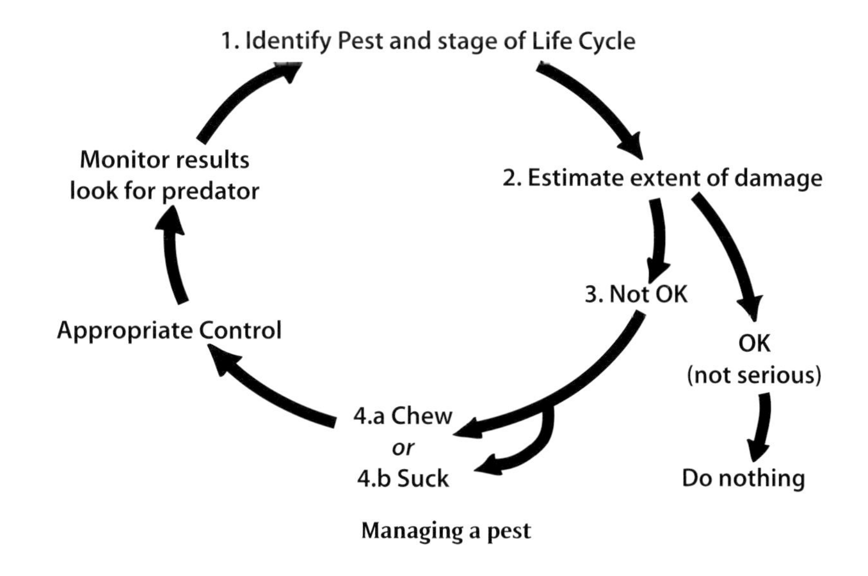

Managing a pest

Last Resort Measures

The use of organically acceptable sprays should be a last resort since they will kill beneficial insects as well.

Derris Dust is a general insecticide so use it strategically. It is made from derris root, a plant. Use on beetles, sawfly, caterpillars, thrips and aphids.

Pyrethrum Derived from the flowers of *Chrysanthemum cinerariaefolium* or *C. roseum*, it is a powerful knockdown poison. It breaks down in about 12 hours and is effective against leaf-hoppers, thrips, whitefly, aphids, lice and fleas. Also kills lizards and bees

Garlic is a general insecticide useful against small and soft bodied pests and won't worry frogs and praying mantis.

Dipel R is a specific insecticide which is only effective against caterpillars, useful if used strategically as it is a biological control. Bacillus thuringiensis.

Nicotine sprays. Don't use, as too toxic altogether.

Bug juices Collect insects, put in blender or mash with mortar and pestle and leave 1-2 days to encourage pathogens and use as a spray. It acts by releasing repellents which tell others to leave and attracts natural enemies or spreads an insect disease. Never use for insects such as mosquitoes or flies which can be carrying human diseases.

Weed brew Fill a 200 litre drum with 6kg of cow manure and add to it six or so species of the dominant weeds in the garden. Let it brew for a couple of weeks then spray it on plants.

Virus spray Place five caterpillars in a bucket of water, stew this up and then spray.

Neem spray is a bitter repellent for grasshoppers and locusts. Can also upset insect hormonal balance.

Fungicides Dilute urine at 1:20-30 parts in water. Milk, powdered or fresh, is a good fungicide at 1:9 water.
Seaweed tea soaked for two weeks in water then washed on. Casuarina needles 1 cup leaves to 1 cup water, boiled 20 minutes, diluted 1:20 for ground spray.

Control principles

Identify the insect and its stage of growth

Become familiar with pest and predators. Carry a hand lens or magnifying glass, check with books. There are two main types of insect life cycles.

Nymph lifecycle Eggs are laid which hatch out as a nymph (stink beetle, bronze orange bug) and these go through many other stages, each larger and stronger than the one before. Eggs are laid in a cluster thus pull off leaf and stamp on it. Last stages are harder to eradicate. Vacuum cleaners have been used quite effectively .

Metamorphosis lifecycle Eggs are laid which hatch out into a grub, maggot, larvae, or caterpillar. This then pupates and later a butterfly or moth appears. The moth stage can lay hundreds of eggs per day, so, try control before the moth has laid the eggs

- Estimate the extent of damage – Can you live with that degree of damage?
- Observe reactions to weather or daylength – Red spider mites increase under conditions of heat and dryness. A light hosing can reduce infestation
- Does it chew? – Caterpillars, snails, slugs. Use stomach poisons, predator insects, hand pick, baits, sprays, coarse substances
- Does it suck? – Aphids, thrips, scale. Use dessicants (salt/flour), predator insects, hose off, vacuum off, asphyxiation (soap, flour, salt)

Special pests

Ask students for their experiences.

Rabbits Two tablespoons of Epsom salts in one litre of water. Dung, tar or kero on/around trunks of trees. Don't like food contaminated with faeces of own species. Don't like smell of mint or rubber (tyres).

Cats Rue is a deterrent.

Nematodes Dilute sugar spray.

Cutworms A matchstick, head down beside each seedling.

Mulch with oak leaves.

White cabbage moth Interplant with sage. Scatter moth balls close to the plants.

Fleas Wormwood trees, pennyroyal, eucalyptus, lavender, thyme oils, diluted and sprayed on. Also must vacuum every five days since seven days is the breeding cycle. Catmint and pennyroyal pillows for cats and dogs.

Ticks and lice Mugwort, Sassafras oil for head lice.

Weevils Bay leaves, cloves.

Bird scare Aluminium foil. Plastic snakes in trees.

Scarab beetles Currawongs and magpies eat these beetles which cause die-back.

Five eucalyptus trees are required for a magpie

family to raise two young.

Tigers in India. Use a scarecrow wrapped in an electric fence.

Student activities

- *Find pest damage, observe it closely and describe what you think may have caused it*
- *Describe three insects and their lifecycles in your garden, and say whether you think they are pests or predators and why*
- *Describe a pest and find out what its predator is and what type of habitat it requires to continue to live in or be attracted to your garden*
- *Investigate one parasite and its host*
- *List three methods of control and state why you would use them*

Extra notes

1. Biological control is the control of one living species by another. In general it is one animal species that controls another. It usually requires food enrichment in the habitat for the predator or parasitoid. There are three types of control by primary parasites:
   - Parasitoid (wasp etc.) lays its eggs on or in another animal (the pest) and its lavae hatch and eat the pest.
   - Predator (bird etc) which lays in wait, or hunts or traps its prey which is the pest.
   - Pathogen a disease micro-organism (bacteria, virus or fungus) which weakens or kills the pest.
   - Secondary parasites are a problem because they destroy the favorable parasite.
2. Interactions in IPM.

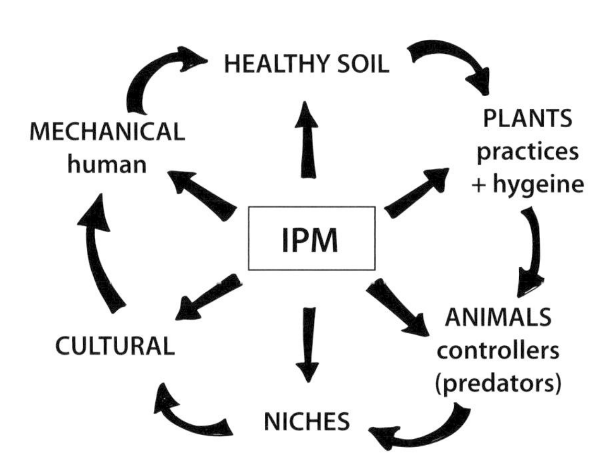

Summary of IMP Elements

UNIT 28
Living with weeds

Weeds have been sadly maligned. They were defined by farmers and decorative gardeners who simply wished to restrict the plants in their gardens and fields to a very narrow spectrum of plants which were largely introduced. Weeds used to mean 'unwanted' plants of cultivated land and particularly farms. Weeds were also seen to compete with economic crops for nutrients and water. They were known sometimes to taint milk, or form burrs in the skin of animals.

Weeds have recently been reviewed by ecologists who have examined the role, structure and value of weeds in natural and cultivated ecosystems. Weeds perform many important functions for the soil. In permaculture weeds are valued for the way they function in designed ecosystems to improve soils, reduce water loss, act as living mulches and provide wild food and nutrients.

Weeds are really the only guardians of the soil because the land selects them, not humans or other animals. In fact, 'weed' is a very good word describing a plant which is a soil protector since its germination and survival is the land's choice. However, weeds can be highly suited to their position and thus the energy input is very high to manage the site so it is no longer advantageous for them.

Weeds tell of soil nutrients, solar exposure, animal grazing stocking rates and other variables such as soil erosion. The land needs more than the current palette of plants that the gardener or farmer is using for the soil. This applies to rainforest, urban wasteland, bushland regeneration, and farm soils tired of animals and/or crops.

Australia has given the world Wild Plant Rescue as a way of saving endangered and plants threatened with extinction a chance to survive. When weeding it is important to check for endangered plants and preserve them.

Learning objectives

In this unit learners will be able to:

- Defend the rights of a weed to exist
- Describe the part it plays in an ecosystem
- Analyse the conditions suitable for its establishment

Teaching tools

- Any sketches can be adapted from the reference books. However, it is strongly recommended that this unit is studied in a place of considerable disturbance and weed infestation
- In Australia that can be urban bushland, in Viet Nam perhaps the hills of tertiary cleared rainforest, and in Cambodia, there are many rural and urban sites

Terms

Climax Community A final or stable community in a successional series that is more or less in equilibrium with existing environmental conditions.

Community All organisms inhabiting a common environment and interacting with one another e.g. Sand dunes, rainforests.

Diverse Having different kinds, forms or character.

Pioneer/Coloniser A plant or animal which successfully invades and becomes established in any ecosystem.

Rampancy Characteristic of plants that grow over others with great vigour.

Succession A slow orderly progression of changes in community composition during development of vegetation in any area, from initial colonisation to the attainment of the climax typical of a particular geographic areas.

Ethics

- Weeds protect soils
- Weed infestations play a role in all ecosystem nutrient cycles
- Weed are indicators of soil nutrient status

Principles

- The presence of weeds act as a guide to soil pH and nutrient status
- There are cogent reasons to control weeds biologically
- Leave weeds for their soil protection and replenishment functions until there is a good replacement plant

Weeds – cause and strategies

Why weeds?

Ask students: Why is a perfectly good plant branded bad?

Branding plants as bad

- Compete with useful plants for water, nutrients and light
- Poison humans and domestic animals
- Taint agricultural produce e.g. Milk
- Harbour disease and insect species
- Harbour vermin animals. e.g. Foxes/rabbits
- Interfere with transport and essential services e.g. Water weeds
- Parasitic – dodder on some crops
- Aesthetics
- Invade or pollute ecosystems

Ask students: What factors favour weeds?

Factors favouring weeds

Weeds thrive in disturbed environments, and with bad management. Weeds gain a hold due to:

- Soft, bare soil
- High nutrient availability
- Moist conditions
- Increased light intensity

Weeds as indicators

- Grazing results in lantana, pattersons curse, thistles
- Fire favours fireweed, bracken, erechthites.
- Chemical soil changes give sedges, sour grass
- Exhausted soils produce bracken, moss, blackberry, thistles.

Weed life cycles

Weeds have the following life cycles and knowing these helps control them:

Annuals: Their entire lifecycle is completed in a single year. Traits are usually small and numerous seeds and fruit which are easily dispersed and highly viable, e.g. Cobblers pegs, winter grass.

Biennials: They normally require two growing seasons to complete their lifecycle. They often have a rosette habit e.g. Dandelions.

Perennials: Plants persisting from year to year and have a variety of habits e.g. Kikuyu, tradescantia, privet and cat's ear.

Areas favouring weed infestation

What areas are likely to be weed infested?
Creeks, dams, roadsides, downslopes, edges, sports ovals, behind houses, railway lines, where animals graze.

Why nature needs weeds

Ask: Advantages of Weeds?

- Loosen soil e.g. Tubers and flat weeds
- Stabilise and protect from desiccation
- Cover as protection from rain, sun and wind erosion, mine and trap minerals
- Change soil nutrient status
- Colonise poor soils
- Absorb surplus water and nutrients
- Part of a succession
- Soil pH can be changed favourably
- Provide a habitat for pest predators

Ask: What weeds do students know that are valuable and what do they tell?

Sorrel: Acid soils
Bracken: Nitrogen depletion
Thistles: High or increased nitrogen levels
Taproot weeds: Depleted topsoils
Bracken, acacias and nettles fix atmospheric nitrogen to help renew soils.

Ecological weed management

The words 'weed control' are simply a euphemism for using herbicides. Chemicals increasingly replace good management of weeds, pests, diseases and fertility. There are three important reasons not to use herbicides:

- Impact on human health
- Impact on the environment
- Impact on gardening and farm management techniques

Ask: How can weeds be managed ecologically?

Most weeds are colonisers, and left to themselves would eventually die out and be succeeded by climax species. For example grasses are succeeded by herbs, and then by nitrogen-fixing species. Generally, it is natural ecosystems that would finally return because climax species are most suited to the combined environmental variables of wind, soil, rain, temperatures and so on. Techniques to manage weeds rely on this ecological knowledge of succession.

First – Leave them until you have more useful plants to replace them or wait until there is a management strategy for the land because bare ground is a disaster!

Second – Plant succession plants amongst the weeds. e.g. Indigenous, short-lived, nitrogen fixing species, then fruit trees and timber trees. Viet Nam has succession fruit trees e.g. Longan and litchi succeed plums and jujube.

Third – Slash frequently, especially seeding annuals and biennials. Compost with hot compost, or make silage, or graze intensively with chickens, pigs etc.

Fourth – Develop a crop such a timber trees, firewood, or orchard species and plan a final climax high income crop.

Fifth – Maintain interplanting with melons or herbal species, or maintain with light grazing such as donkeys, geese, goats (good fencing), pigs or chickens.

Sixth – Study local weed biology so as to know when seed sets, dies back under frost, etc. and handle other farm operations to with the weed management.

Other techniques

- Drown by flooding
- Weaken weeds by restricting light
- Change pH status
- Use rampant species to choke other species.
- Attract weed diseases
- Mulch heavily
- Burn with hot fire
- Change land use and so on...

Legal Controls

Ask students for the difference between a community pest and a noxious weed.

Where there is gazetted community awareness, a plant is a community pest e.g. Pampas grass, it is a noxious pest where a nursery is not permitted to sell the plant and the Council and residents must eradicate them from their own land.

In New South Wales, SEPP, State Environment Planning Policy number 19 (1986) deals with Bushland in Urban areas and weed infestation.

Student activities

- *Find the most common weeds occurring near your place, describe the weed. Describe its habitat. Why does it grow there? How would you control it?*
- *Find a piece of weed infested land and carry out weed control over about one year – you need to watch a plant over different seasons*

UNIT 29
Aquaculture

This unit links up with those on Water and Designs in Nature.

Fish farming is an ancient practice, there is nothing new about it. More than 4,000 years ago in China, farmers were feeding carp on silkworm wastes. Early Roman writers, including Pliny, gave directions for the use of stable sweepings and other organic wastes in fish ponds. In Europe, fish farming developed along with the monasteries.

An aquatic environment can trap and store much more of the solar radiation falling on it than can the same area of good grassland. Converting this primary production into fish flesh is one of the most efficient methods of obtaining high quality animal protein.

Aquaculture is a fastest growing form of land use and is best integrated with land systems. No one system can utilise everything, so the polyculture aspect is important. Aquaculture can yield between 250-1,150 kilos of protein per hectare.

Aquaculture is a safety feature in any design since it is the water component which is so critical in wet climates where it assists in drainage, or dry climates where it is a water resource. The water can be used as Second Class water for animals and garden and possibly even for washing clothes. It also modifies climate. The Vietnamese understand this very well indeed and the Fishpond, or 'Ao', is a fundamental part of every design.

From Wikipedia, the free encyclopedia:

> *Aquaculture, also known as aquafarming, is the farming of aquatic organisms such as fish, crustaceans, molluscs and aquatic plants. Aquaculture involves cultivating freshwater and salt-water populations under controlled conditions, and can be contrasted with commercial fishing, which is the harvesting of wild fish. Mariculture refers to aquaculture practiced in marine environments.*
>
> *The reported output from global aquaculture operations would supply one half of the fish and shellfish that is directly consumed by humans; there are issues about the reliability of the reported figures. Further, in current aquaculture practice, products from several pounds of wild fish are used to produce one pound of a piscivorous fish like salmon.*
>
> *Particular kinds of aquaculture include fish farming, shrimp farming, oyster farming, algaculture (such as seaweed farming), and the cultivation of ornamental fish. Particular methods include*

aquaponics and Integrated multi-trophic aquaculture, both of which integrate fish farming and plant farming.

Learning objectives

By the end of this unit students will be able to:

- State why aquaculture is highly efficient
- List factors required in setting up large and small systems
- List criteria for fish selection
- Establish urban aquaculture

Elective:

- Give relevant stocking recommendations
- Describe aquaculture management
- Describe and setup an aquaponic systems

Teaching tools

- Visit an aquaculture nursery
- A garden with small ponds
- Photos of water systems

Terms

Aquaculture Complex wetlands, water environments, yielding plant and animal products.

Aquaponics Usually small scale fish, eutrophic water and plant cultivated ecosystem suitable for home use and highly productive and produces no waste.

Fish farming Growing fish species for consumption and usually a monoculture maintained by high energy inputs.

Mariculture A marine complex of saltwater plant and animal species.

Trophic level Each link in the food chain is called a trophic level.

Ethics

Store water on land, use it to stretch out dry periods, and to meet some household water requirements, and clean it before it leaves your land. A store of water like this can also provide a multitude of products, functions and microclimates.

Principles

- Aquaculture is one of the most efficient methods of obtaining high quality animal protein
- Fish are cold-blooded so they don't use energy to control body temperature
- Fish can devote more food energy to growth than land animals because their weight is supported by water
- Fish can be grown on wastage such as animal residues
- Fish farming can be carried out on marginal land so making it productive
- Fish ponds can add another use to existing facilities such as irrigation dams and can occasionally be dredged for fertiliser, thus complementing agriculture and increasing overall food productions
- Fish farming shortens the food chain especially when species which feed on low trophic levels are chosen e.g. Yabbies, herbivorous fish, mussels

Cambodian Aquaculture

Fish pond ecology

Each species of fish feeds on a defined array of micro-organisms.

Each species alone cannot utilize all available food, e.g. An edge species such as a mulberry provides much fruit which is eaten by fish in water and on land by ducks.

The ducks' manure in the water is used as food by fish and by micro-organisms which feed fish.

The mulberry also feeds insects which drop into the water and are eaten by fish, as is the frass.

Leaves which fall in the water are also eaten, especially by shrimp.

Use a Diagram for the above – IT'S A FOOD WEB!

When the food is not fully used, imbalances develop with a drop in oxygen levels.

Yields

In terms of protein, four hectares of water is more productive than 80ha of grazing. The nutrient water from fish and animals is very rich. Generally, water from an aquaculture has an excellent pH level and is best used for tree crops. Sometimes there is too much nitrogen for vegetables.

Factors in setting up

pH Best between 7 and 9. Anything under pH 6.5 won't be very productive. pH 7.5-8.0 is optimum but will change at different times of the year.

Depth At least 2m deep in the middle so fish can escape during heat of summer. More than 2 metres is not valuable. The ideal temperature is between 18-25°C. Trout will start to die if water temperature reaches 29°C.

Clear water Hold a 20p coin 18" under the water – if it can be seen clearly the water is clear enough for fish. Turbidity in water reduces the sun's penetration and thus algal growth. A small amount of obscurity can inhibit predators. Gypsum will clear muddiness and help balance the pH – 560kg per ha added in small quantities is effective in clearing murkiness.

Food Fish farming is much more economical if you grow your own food. A new dam can be inoculated by taking a bucket of water from one dam and putting it into the new one to breed up plants and micro- organisms. Algal feeders need heavy manuring and algae needs sun. Grass eaters need water plants along the edges. Taro is an excellent feed with weeds in between.

Mulch the edge of a new dam immediately to encourage edge plants and reduce erosion and run-off. Suspend nets vertically in water to grow algae. The algae, Macrobrachean rosenbergei, will grow as far south as Coffs Harbour.

Existing vegetation Must be assisted to attract insects or drop feed into the dam. It also provides an immediate habitat for micro-organisms.

Carrying capacity Size, number of fish and carrying capacity is related to the surface area, and not to the depth of water or total volume.

Edge Water level stability and water loss

It may be necessary to ensure that the dam can be topped up if required. It is good to have a few water tests made. Oxygen levels are one of the most important factors. A reticulation system or flow forms can be useful.

Design of dams for aquaculture

Shape The longer the edge, the greater the shallow area, the more food is available.

Depth Different depths are important in providing a range of different habitats; large to the deeper water and deep water will ensure a region of lower temperature during the summer months. Shallow water can support weeds which offer protection to small fish, and are a source of large quantities of food. Shallow water plants provide habitat for waterfowl, and produce food crops such as water chestnuts, taro, and arrowhead which comprise the polyculture.

Drainage allows proper management practices, e.g. Cleaning of the pond, removal of sludge for use as fertiliser, and complete harvesting of the fish if needed. A screen at the overflow is essential to allow for drainage. Have bottom sloping towards the outlet.

Catchment area is managed in complementary ways, and vegetated to reduce muddiness of water. No sprays to be used and stock should be excluded.

Shelter, such as tyres, terracotta pipes and old logs need to be provided.

Pond bottom is very important in the biology of the body of water. A good bottom is able to quickly recycle nutrients and make them available. On a poor bottom, decay is slow. Gravel, clay and sand bottoms can be improved by the addition of organic matter such as stable manure, sewerage sludge or by sowing a green manure crop before filling the dam with water.

Criteria for selection of plant and animal species

Pond size See accompanying table. There is a surface area, edge and depth suitable for certain species and which affects their stocking rate and food supply.

Climate Water temperature, maximum and minimum temperatures and overall geographic areas e.g. Inland, coast or mountains, need to be considered.

Available sunlight (In the southern hemisphere) plant large species on the south side of a dam since large trees on the north shade sunlight when it is usually most important.

Evaporation Temperature, wind speed and rainfall all interact. Water levels may need to be topped up.

Environmental impact Whether species can escape and become pests, the interactions between species, and symbiosis are all important considerations. Something needs to be known of the food chains in the water, and their inter-relationships. Take note of government regulations – fish stocked in dams should reflect the natural population of the areas e.g. Murray cod, golden perch and silver perch are found naturally only in the Murray-Darling basin. Catfish and Australian bass are preferred for the eastern watershed.

The more suited the fish to their habitat, the faster their growth rate.

Wind Breezes re-oxygenate the water. Appropriate wind machines can be used to oxygenate water, also perhaps, plant/wind funnels, see Windbreaks.

Water Quality Depends on the amount of sediment, the water-shed and pollution.

Before stocking A new dam should be allowed to settle for at least three months to allow for the establishment of a good food supply.

Management

Recommended fish stocking rates

Perimeter (metres)	Approx area (hectares)	Silver Perch Catfish	Golden Perch Aust Bass	Murray Cod
100	0.04	25	25	25
200	0.22	75	50	200
300	0.54	200	125	75
375	1.00	350	125	75

Source: NSW Dept of Agriculture

1. Another calculation is one mature fish to 15 metres of edge or, one mature fish to 15 square metres of surface.
   When stocking fingerlings there will be some losses.
2. Do not stock low, the fish grow very big and are hard to catch. With very high stocking rates, the fish stay small.
3. When newly stocking check that there are no other fish or eels. Eels can be trapped by adding Derris Dust to empty dams which asphyxiates them. After two weeks the effects of the Derris Dust are gone.
4. Do not stock inland dams with coastal fish and vice versa.

Species

- Murray Cod – 200 fry per hectare; 120 fingerlings/ha. in cages. They like to hide under floating rafts, and in logs and clay pipes.
- Golden Perch – 300fry/ha – they are tiny, tiny.
- Silver Perch – 160 fingerlings/ha.
- Catfish are natives and dam breeders and very good eating. Red Fin is a good small breed if it cannot escape.
- White Amure is a fine eating fish. It is a plankton feeder. Prawns are eaten by big frogs; goldfish eat mosquitoes.
- Plankton Eaters; i.e. rainbow trout, can only breed in running water – through flow forms.

Small fish species

- Water snails
- Damselfly nymphs
- Small forage fish
- Carnivorous fish

Feeding Fish

- Don't supplement as the cost is too high, use homegrown feed
- Add one thing at a time to the water and observe what happens
- Edge plantings of trees attractive to insects e.g. Mulberries, bottlebrush, and tea-tree
- Use sweet flavours to attract fruit flies, use honey and sugar. 60% of fish food is insects. Can use a brick boiled in liver to attract the blowfly family
- Fertilising with manures encourages algae growth and this allows denser stocking rates
- Nutrient Cycles

Nutrients

Phytoplankton
Fixed aquatic plants
Animal plankton

Harvesting

- Harvest smaller/medium sized fish
- Use traps or line to catch fish – nets are illegal
- Fly Traps (meat waste), light traps (solar-power) can be used

Lack of Oxygen

Occurs during hot weather (warm water contains less oxygen than cold water). May occur after rain when organic matter such as animal manures, vegetable matter has been washed into the dam. Decomposition of organic matter uses up available oxygen.

Signs of oxygen deficiency are dead fish or, fish coming to the surface gasping for air. Oxygen may be replaced by circulating the water, or pumping it up and spraying it back onto the surface.

It can be avoided by maintaining a better balance in the first place.

A windmill – paddle floating on dam. Air rippling on water will work. Ducks swimming on the water. Solar pumps.

Predators

Cormorants are the major bird predator.

Their visits are infrequent and irregular but once a dam with fish is found, they work the dam until the majority of the fish are taken. Fish quickly learn about safe retreats. Earthenware pipes, hollow logs, old tyres tied together and plastic pipes can be used, but not metal pipes or chicken wire as they release chemicals into the water.

An abundance of forage – fish or crustaceans (shrimps, yabbies, goldfish) – will ease the predation pressures on the dam fish.

Undesirable fish

Eels are very common in farm dams on the coastal side of the ranges.

They eat fingerlings and so significantly reduce the chances of establishing fish in a dam.

They can be removed by a few nights trapping, using lights and baits of fresh.

Pollution

Fish waste (when there is only one species) can build to levels that foul the water and which inhibit growth.

Overcome by stocking several species to match different levels of available food.

Design strategies

Use screens between predator and non-predator fish.

Rafts with suspended nets separate predator fish from others.

Predator fish only get the little fish that swim by.

Yabbies like to hide in little things. They will dig holes in dam walls. They love living in beer cans (non-rust aluminium) suspended from a raft. Yabbies will eat worms. In marshes and wetlands make sinks to grow fish, yabbies and prawns. This can also be done in mangroves.

Freshwater Mussels; can be grown on ropes and can filter 200 gallons of water per day (cleansing like kidneys). They also deposit phosphate.

A main pond with other ponds around it is an option.

Water Plants

Gradual shelving from 'dry' land to 1.2m deep. If the water levels cannot be topped up then have a floating raft and place the plants in it. If the raft is too high, it can be weighted down with baskets of water lilies. Just add soil until the raft floats at the right level.

Plants should include those that prefer:

- Running water
- Grow on edges
- Still water
- Edible roots
- High nutrition value
- Economic products
- Edible fruits

Taro, Chinese water chestnut, duck potato, bullrush, waterlily, lotus, Indian water chestnut, *sagitarria*, cumbungi. Watercress prefers running water. Vietnamese cress will grow on land or water.

Kang Kong grows 0.1-0.2m with duck potato. It is sometimes called swamp lettuce and is very high in nutrients and mineral value.

Water chestnuts have an excellent market prospect – one square metre will grow from one quorm. They should be harvested as clean as possible. In a sub-tropical climate they require 8-9 months to grow, then given a top dressing of manure for next year's growth. They die back naturally in cool weather. They too like a high pH. Divide when harvesting.

Bullrushes – the whole plant is edible and the roots can be eaten like potato.

Trees on the edges

Mulberries, bottlebrush to attract insects, fruit trees, avocado (on the south edge where the micro-climate can be 4-5°C higher than the surrounding area), quinces and whatever grows locally in wet areas.

Shrubs and herbs on the edges

Comfrey, sweet potato, lavenders, lemon grass, fragrant plants, millet, passionfruit, kiwi fruit and tea-trees.

Diagram of shelving edge showing a large range of plant species.

Good aquaculture enterprise combinations

- Combinations of fish/worm farms, or, fish/pig or, fish/ducks
- Ducks and Fish. In Hungary, 500-600 ducks are used per hectare
- Fish and Pigs Basis of VAC system in Viet Nam
- Fish and Domestic Waste
- Fish and Agricultural Waste e.g. Rice hulls, crayfish
- Fish and Industrial Waste e.g. Abattoirs, sugarbeet processing
- Fish and Worm Farming

When up to six species of fish and waterfowl are stocked, the predators of fish take no more than 15% of fish. Herbivorous fish perform a special function, the Chinese say, "if you feed one grass carp well, you feed three other fish". Grass carp consume massive quantities of vegetation (own weight in a day), and excrete large quantities of partially digested materials which directly feed bottom-feeding fish i.e. common carp, and stimulate production in other parts of the food web. Grass carp can grow as much as 3-4kg. per annum. (The Chinese use mulberries particularly since it feeds duck and fish on fruit and the leaves feed shrimp and grass carp).

Illinois channel catfish fed on commercial pellents have a maximum production of 1,500kg/ha. With manuring and stocking of Chinese carp, production increased to 4,585kg/ha.

Production can be increased three times with pig manure/sun/carp. Then, water plants growth is increased with the nutrients from increased fish stocking and growth.

A successful polyculture has a mixture of fish, crayfish, plants, molluscs, water fowl and edge plants.

Seepage areas can be used for mints, bamboo, and trees such as willows, pecans and poplars.

Changing enterprises

If desired, after a few years the dam can be drained and terraces of sleepers etc. are put in place and this will be an excellent planting area for almost any crops.

Multiple Pond Systems

These are for a large grower of fish. Multiple ponds are set up and minerals like lime can be added to higher, smaller dams and allowed to trickle down into the lower ones.

Urban water ecosystems

- Low maintenance
- Don't need watering, weeding or mulching
- Attracts birds to the garden
- Raises the humidity of the air around the pond
- Good for subtropical plants such as paw-paw
- Provide a habitat for insect predators such as frogs and lizards
- Plants themselves are very ornamental

Siting

Ponds need full sun and a low site looks more natural.

Containers

A pond isn't necessary, old concrete laundry tubs, baths, fishtanks can be used, or a recycled tractor tyre turns into a pond.

Plants in containers are advantageous because water is clearer, they allow for easy harvesting of plants, easy repotting and division. The pond is easily cleaned. Use a clayey soil with compost or well-rotted manure.

Requirements

Water proof lining.

Submerged oxygenating plants such as tape grass, water milfoil, water, thyme etc.

Scavengers help to establish the natural balance, fish and water snails clean up rotting vegetation and algae, goldfish eat mosquito larvae and other insects.

Plants and fish prefer mature water so do not empty the pond unnecessarily and top up gradually.

Fertiliser

Can be small amounts of compost or manure.

Oxygenation

Water lilies serve a practical function by keeping oxygen in water by trapping it under lily pads.

Aquaponics

From Wikipedia, the free encyclopedia:

A small, portable aquaponics system

Aquaponics is a sustainable food production system that combines a traditional aquaculture (raising aquatic animals such as snails, fish, crayfish or prawns in tanks) with hydroponics (cultivating plants in water) in a symbiotic environment. In a aquaculture, effluents accumulate in the water, increasing toxicity for the fish. This water is led to a hydroponic system where the by-products from the aquaculture are filtered out by the plants as vital nutrients, after which the cleansed water is recirculated back to the animals. The term aquaponics is a portmanteau of the terms aquaculture and hydroponic.

Aquaponic systems vary in size from small indoor or outdoor units to large commercial units, using the same technology. The systems usually contain fresh water, but salt water systems are plausible depending on the type of aquatic animal and which plants. Aquaponic science may still be considered to be at an early stage.

Student activities

Design a tyre-pond for a garden and give reasons for

- *Site selection*
- *Plants species chosen*
- *Anticipated yields and advantages.*

UNIT 30
Wild friends

People are often in conflict with wildlife. The human species is one of the few animals not threatened with extinction from reduced numbers. Wildlife is more in danger from people than people are from wildlife.

On the whole wildlife loses, except perhaps for rats, cockroaches and pigeons (often considered vermin). With care and understanding people and wildlife can live together, or at least co-exist.

Many growers find themselves in conflict with wildlife. Yet what is required is a balance, having wildlife on land and around living areas, without severe raids on food crops.

This unit complements Unit 25 – Integrated Pest Management, Unit 22 – Zone V and Unit 23 – Weeds.

Ecologists are just beginning to understand how important each animal is to its ecosystem. For example, the removal of a small apparently useless shrub in Australian bush to make more room for pasture meant the death of a wasp which preyed on a devastating grub. In Japan, sprays used on plum trees killed the bee pollinators and there were no fruit for many years. Some tiny insects are keystone species which means that many other species depend on them even though they are very small and do not seem of economic importance. So our primary aim is to deter and not to kill.

Learning objectives

Students will learn to:

- Encourage wildlife for the assistance they give
- State what benefits wildlife bring to land
- Suggest means to keep raider animals in check
- Design habitats that are animal friendly

Teaching tools

- Design their own as appropriate for their own bioregion
- Students to draw on the boards, or show types of deterrents
- Good diagrams of wildlife corridors which link up farms, villages and national parks are useful

Terms

Decoys Food or attractants placed to attract animals to another crop.

Insectivorous Having a preference for eating insects.

Ethic

The basic psychological and spiritual position that all life has 'right to life' is a valuable starting place.

Principles

For wildlife our goal is for them to carry out the following valuable functions:

- Pollination
- Predation
- Sanctuary
- Seed germination
- Grazing, browsing
- Beauty and wonder
- Cleaning
- Tractoring
- Diversity

Deterring animals

The principle of deterrence rather than killing is now well accepted and helpful research is being carried out along these lines.

Use Decoys e.g. Sow turnips, radishes or buckwheat around grain field to decoy rabbits. Scatter grain around orchards to keep birds in large numbers from attacking fruit.

Use Deterrents Derris root, mint and rubber tyres are said to repel rabbits.

Unpalatable crops – recent work in UK on flavours birds dislike is quite effective – these are bred into the fruit.

Silhouettes of hawks to scare birds (must be removed immediately the crop is harvested so birds do not get too used to it). Scare guns, flags, and singing wires are also effective.

Some people tell raider animals very firmly what is theirs and to stay away from the rest. Mechanical barriers – for full details of these, see unit on Integrated Pest Management.

Electric fences for cattle/pigs.

Angled fences for kangaroos.

Snakes: dig pits to capture them. Or place gravel paths which they don't like crawling over.

Plastic snakes in fruit trees deter birds if they are changed round in the trees every two days or so.

Tigers: place an electrified human dummy.

Encouraging animals

Wildlife Corridors These are of vital importance,

hence the great significance of roadside vegetation; of planting projects which connect small reserves with National Parks; of corridors on farms along rivers, creeks and farm roadways. Often this is the only habitat some species have left (see reference Corridors for Conservation). Remember they buffer against drought, fire and flood.

Water Birds – only 26% of Australia's birds are nomadic so the rest must have a constant supply of water. Often a variety of water is needed, from deep to shallow; from still to running; exposed and with overhanging vegetation; some birds need to drink several times a day or wash to keep clean and cool. Waterbirds are migratory. They require a hectare of surface water to be attracted to an area.

Create habitat Around dams/ponds/rivers/lakes put dead logs in water, rocks on the side, shelving sand, reeds for breeding birds. Design dams that back into forest or create forest around dams so that animals coming to drink feel safe e.g. Wallabies. The Franciscan nuns at Stroud put in a dam in woodland especially for wildlife, one third of their land is dedicated inalienably to wildlife.

Vegetation

- Leave old trees – often sugar gliders and kookaburras can only nest in old, dead or half-dead trees
- Plant special insect-eaten trees that insect-eating birds will be attracted to e.g. *Acacia melanoxylon* as used by Fukuoka
- Ensure an under-canopy for small birds to live in as magpies and currawongs are tall-tree nesters and prey on small birds
- Plant special trees for special species e.g. Koalas, flying foxes
- Use spiny and prickly plants to protect small birds and other animals from predators especially cats

Food for wildlife This is a little contentious since some authorities in countries such as Canada and increasingly Australia, are asking people not to feed wildlife because:

- Tameness makes them vulnerable to ruthless people
- Wrong foods kill and sicken animals which have a tendency to eat exotic foods
- They can build up dependency on being fed

However, there are some foods which can be offered to attract animals which are threatened if land is cleared land etc.. These are:

- Bran for wallabies
- Grass lawn for kangaroos
- Zoo prescription for feeding birds
- Wild berries and fruit for birds

The best way to attract wildlife is to create habitat.

Student activities

- *Make an inventory of animals seen in their garden or in their area*
- *Do a similar one, after a year when their garden is well-established*
- *Design special animal attracting features in their garden/land. What animals will be attracted by these?*
- *Work out which animals you want in your garden and how to encourage them*

Extra information – Australian bias

Kookaburras should only be fed meat with hair and bones on, as straight steak etc. upsets their digestion. Hence small mice, lizards and rats are acceptable. Kookaburras will also control field mice and rats around homesites.

Offer grated cheese and fat in winter to small birds such as grey thrush, which will in turn eat caterpillars and slugs. They require shallow, hollow logs in stumps to nest in.

Attract insects using yellow colours in the garden and robins, honeyeaters and grey thrush will be attracted to the area to eat the insects.

Pineapple sage and grevilleas growing in orchards will bring honeyeaters.

Spray 'Biodynamic 500' outside cropping fences and kangaroos and wallabies will come to graze.

Wallabies can be encouraged with pollard and ground oats, as can wombats.

Echidnas like milk and a deep litter mulch in hedges and bushland.

With new plantings, there are often no nesting places for animals, so encourage wild ducks by placing five gallon drums open at one end near water.

Lots of stones will bring a range of lizards to eat snails and slugs.

Tadpoles like small and larger ponds of still water.

Lyre-birds and brush turkeys need deep litter and mulch to scratch around in.

Recipe for Feeding Native Birds

Don't feed sugar and bread to parrots as it causes runners – wingless young which hatch and never develop wings.

Don't leave honey out for birds for long – it can

ferment and spread disease amongst them.
Plant nectar and fruit trees and shrubs.

Dry Mix

3 cups of baby cereal
1 cup rice flour
1 cup glucose powder
1 cup egg plus biscuit mixture (from pet shops)
Optional – 1 teaspoon multi-vitamin mixture (from pet shop)
Often a half orange, or piece of apple or other fruit is good for animals.
Some animals, like humans, will 'pig-out' on certain foods which can damage them, e.g. Sunflower seeds.
Make the greater effort and try to design habitat.

SECTION 6

Social permaculture

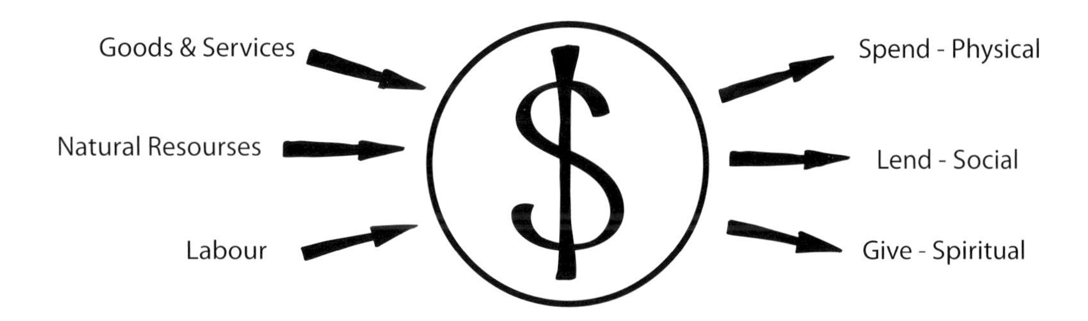

UNIT 31
Bioregions-collective self-sufficiency

Students naturally understand the many invisible laws and structures which operate on their lives. They all experience such things as increase in the price of rice, or an increase in house mortgage interest rates. Many people feel powerless against them. This unit and those following help students understand how invisible structures work and assist in making better decisions about their lives and communities.

A bioregion is an association of residents of a natural and definable region. It can be defined by roads, water e.g. Ocean, rivers, languages or common interests. They can also be physical, such as islands and mountain tops. Classical western models of Bioregions are interesting communities such as principalities, convents or armies

Tribal peoples often defined bioregions well. Tribal maps for clans, totem groups or tribes often select a particular tree or animal, itself limited in distribution, by the sum of topographic and climatic factors.

The acid test of a bioregion is that it is recognised as such by its inhabitants. Populations of bioregions are from 4-7,000 people and can start with as few as 100. Remember, about half will not be active.

Increasingly cities and towns are carrying out bioregional audits to estimate future needs for food and water. Seattle has a very good one. Liz O'Connor of Katoomba in the Blue Mountains of NSW did one for this town. In early 2012 the local government area of Carey in Victoria, Australia carried out one and found a surprisingly high degree of food insecurity.

To some extent bioregions overlap with the transition town movement. Transition towns grew out of permaculture. Many of the same goals are shared. Perhaps transition towns are narrower in scope and focus than bioregions.

Essentially the bioregional goal is for local independence in finance, produce, work and land use. All these require inbuilt resilience to pertubation.

Learning objectives

By the end of this unit, students will be able to:

- Describe a bioregion and develop ethics for it
- Set up a bioregional directory
- Name bioregional services and resources and identify opportunities

Teaching tools

- Tables of inventories which strengthen bioregions
- Charts for students to fill in with you

Terms

CES Community Exchange System. An online exchange of goods and services to meet human needs.

Freecycle A local community online network which offers good and service for no charge.

LETS Local Exchange and Trading System which is a formal community system for exchanging or trading goods or services using local currency not federal.

Ethics

Each bioregion requires a clear ethical basis in its charter. The Haudenousaunee (Iriquois) Indians, for example, considered the advantages and disadvantages of any decisions they made on their descendants for the next seven generations.

Bioregional ethics are directed at resilience, self-sufficiency in products, and culture; with sustainability in soils, water, biodiversity and,replenishing these. Diminish none of these.

Explore these ideas with students. Revise the difference between sustainable and self-sufficient. Why is it difficult for individuals to achieve self-sufficiency? What would happen to global trade and resources if bioregions operated effectively?

Ask students to draft bioregional ethics for where they live eg preserve natural character or improve the sustainability of the region.

Principles

- The priority is for bioregional stability
- Consider most exports and imports as impoverishing
- Bioregions need to be political and financial units with fast communication systems
- Develop a network of organisations with everyone in the community having a link into a bioregional organisation

The essence of sustainability lies in the bioregional context where self-sufficiency can be achieved.

Local purchasing

From Wikipedia, the free encyclopedia:

> *Local purchasing is a preference for buying locally produced goods and services over those produced more distantly. It is very often abbreviated as a positive goal 'buy local' to parallel the phrase think globally, act locally common in green politics.*
>
> *On the national level, the equivalent of local purchasing is import substitution, the deliberate industrial policy or agricultural policy of replacing goods or services produced on the far side of a national border with those produced on the near side, i.e. in the same country or trade bloc.*

Local economy theorist, Michael Shuman, sums up local economy as a tension between 'TINA' (There Is No Alternative), and 'LOIS', (Locally Owned Import Substitution).*

Compile an ethical directory

Bioregional organising begins by carrying out an inventory or audit or directory to see what is there to work with and, what is missing. It requires a group of people to:

- List by headings all the primary needs of a community or society
- Write criteria for each of these and then write the actual goods and services
- List all groups, individuals and organizations with compatible ethics for goods and services.
- Then indicate whether they meet the needs of the bioregional population or are absent

By listing what is available and unavailable the directory highlights the untapped resources and markets. Bioregional enterprises can then be researched to start to fill the gaps.

+ indicates the product is locally available, or could be

– indicates it is imported or not available locally.

The inventory, which is also a directory, has been started below.

Complete each list with all the names of people and their services for local human and animal needs. Also included are environmentally sensitive areas not covered by protective legislation. Don't include the 'need' for luxury wants.

See the final chapter in the *Designers' Manual.*

Ask students to list their ideas of criteria for: Finance, leisure areas, employment and support service, transport systems, social security programmes.

Priorities for building bioregions

- Look at leaks in finances and resources and plan to block them
- Start a newsletter
- Form producer organisations, agricultural/food gardens/markets
- Care for the welfare needs of your bioregion.
- Run courses and seminars for sustainability and resilience e.g. Every household a garden, or every household to have chickens
- Produce some luxury items
- Keep historical records. A community without history is like a person without money.

Fill the gaps by offering public information about:

- LETS systems
- Building

* *The Small-Mart Revolution: How Local Businesses Are Beating the Global Competition*, Michael H Shuman; Berrett-Koehler Publishers, 2007

Food	Shelter	Energy	Finance	Transport	Water
Local Organic Indigenous Economic Good nutrition	Water and energy efficient Local materials Local workers	Renewable Local Non toxic	Local Ethical LETS Markets Freecycle	Accessible roads for bikes, foot traffic Electric Small scale	Regulated Cleaned Stored
Drinks Teas +, Coffees – , Fruit +, Alcoholic +, Milk +, Soy – , Dairy –, etc. **Sugars** Honey, Cane, Beet, etc. **Staples** Potatoes, Rice, Flours, etc.	Timber Mud Hay Strawbales etc.	Renewable Solar Wind Wave Geothermal Wood Bricks etc.	Grower markets CSA Swaps Garage sales Food co-op etc.	Bicycle hire Local free Bus Trains Car share Car hire etc.	Tanks Springs Rivers Lakes Dams etc.

- Hold courses and workshops
- Build co-operation between suppliers and consumers
- Start a CES on line
- Open a local freecycle online
- Ideas for energy saving, renovating
- Local services
- Ethically sound organisations

Ask each student to write down what they think the priorities are for their bioregion. Then list workshops required to meet needs.

The local economy

Run two economies:
Informal Economy (barter and no money):
- Work groups to achieve projects
- Barter groups to exchange goods and services
- Informal bartering

Volunteer labour to individuals or groups:
- Formal Economy (uses money)
- Consumer-Producer co-ops
- Community savings and loans
- Bioregional currency systems
- Leasing systems

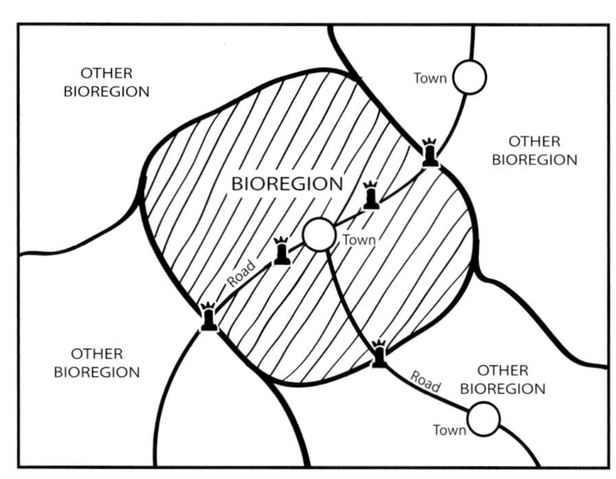

Indian Bioregional Tax System

Students work in groups to list the advantages and disadvantages of each type of Economy i.e. the informal and the formal.

Student activities

- *Describe the characteristics of their bioregion and write ethics for it*
- *List five sustainable activities in their bioregion*

UNIT 32
Ethical money

Economics should be a branch of ecology. Every enterprise needs costing for its destructive as well as economic results, e.g. Deforestation results not only in timber and paper-pulp production, but also in soil and nutrient loss, river silting, erosion etc. and these effects are not costed in conventional economics.

Money is a communications system. It cannot be separated from people and environmental issues.

At one time most major religions and cultures banned interest on loans. Islam still has an equity system. In Europe, Christians were not allowed to charge interest on money. Jews were allowed to charge interest only to non-Jews. Before banks began charging high interest rates many people spent their additional money on arts and crafts, i.e. investments in right livelihood.

Interest is the main money evil. Earning money from interest returns nothing to the country's wealth unless it is ethically invested. The Rothschild empire earned four times the GNP of Germany between the wars on interest alone. Money only works in a growth economy.

Every amount you pay and everything you purchase includes interest on capital e.g. Electricity and water bills include interest on depreciation, or on new capital works such as dams and power-houses. If you borrow money at 12% the capital owed will double in six years. If you borrow money at 20% it trebles in five years.

Learning objectives

By the end of this unit, students will be able to:
- Give examples of non-formal and formal economics
- State three types of socially responsible investment
- Name two community economic systems.
- Write a list of their assets

Terms

Formal economics Economic transactions subject to auditing procedures.
Informal economics Economic transactions not subject to auditing procedures
Macro-economics Transactions affecting society
Micro-economics The personal effects of one's own transactions.

Socially Responsible Investment (SRI) That which creates sustainable and resilient wealth and natural systems.

Ethics

Money used in socially responsible ways enriches a bioregion. However, human needs are not self-evident nor necessarily easy to meet.

Principles

- Money has a multiplier effect within a community.
- Generative and procreative investment increase community prosperity.
- Question banks and their function. Wealth can be measured as biological richness, good health, productive work and is not restricted to money.
- Interest gained by large organisations largely flees the community and is dead money. Interest placed in Socially Responsible Investment (SRI) returns to the community.

'Socially Responsible Investment' (SRI) defined

An investment is considered socially responsible because of the nature of the business the company conducts. Common themes for socially responsible investments include avoiding investment in companies that produce or sell addictive substances (like alcohol, gambling and tobacco) and seeking out companies engaged in environmental sustainability and alternative energy/clean technology efforts. Socially responsible investments can be made in individual companies or through a socially conscious mutual fund or exchange-traded fund (ETF).

Read more: www.investopedia.com/terms/s/sri.asp#ixzz1pvTRo4Ql

Multiplier effect of money

Acquiring Money:

- There are three ways to get money in
- There are three ways to use it

Goods and Services	Spend-Physical
Natural Resources	Lend-Social
Labour	Give-Spiritual

Retaining Money

When money is brought into the community and circulates, it increases the wealth of the community. As soon as money escapes the community, there is the loss of wealth. As little money as possible should flee the community.

The implications are to use local services and to buy and sell locally.

Pass $1.00 around the class and each person receives something for that dollar as the money circulates. The group is $15 now better off in value. As the $1.00 circulates the second time, privately ask one person to pocket the $1.00 and leave the room. The group is now unable to trade. It is poor.

Unemployment benefits ('sit-down money' in Aboriginal language) can go much further if people put it into one account. e.g.

1/3 – Local investment i.e. Food co-ops
1/3 – Cash uses
1/3 – Local banking.

Community and personal wealth

Reclassify personal wealth according to use and renewal of natural resources. Considering biological multiplication of resources leads to the following way of naming assets.

- Degenerative assets deplete non-renewable resources, and are those that decay, rust, or wear out and usually pollute. They are

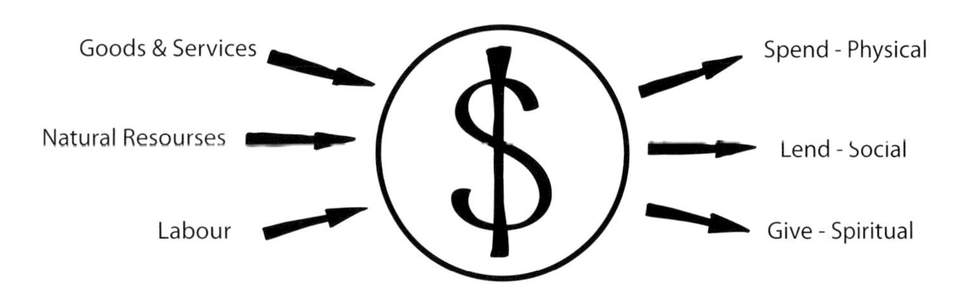

Money In and Out

buildings, cars, machines, computors etc. In communities most public money is spent on degenerative things. Banks lend mainly for degenerative items and call it infrastructure e.g. Huge dams

- Generative assets manufacture or process raw materials into useful products, e.g. Grinders, blenders, lathes
- Procreative assets multiply over time e.g. Trees, animals systems such as pig tractors or pig litters
- Informational assets are ethical knowledge in news, films, books, people etc
- Conservational assets conserve resources e.g. Trees help to clean and conserve water and soil

---

Ask students to consider their assets. Make a list under these headings and see if they can shift their assets wealth from 1 to 3.

Militarism

Australia spends $25m per day on its defences and it is currently not at war. Compare this with housing, education, welfare, environment or foreign aid budgets e.g. Australia gives $40m/yr to the Asian Development Bank and then complains about economic refugees. Clearly the scales are totally out of balance.

The usual curve for growth for things like roads, hospitals, schools, housing is like that in the diagram below where eventually everyone can have enough for their needs.

Militarism is the greatest waster of human money, money that is needed for socially responsible investment. The military graph has no period of maturity, or senescence. The bill is limitless.

Banking

Money placed in bank accounts is only a loan to a bank. People often give up all responsibility for their money once it is deposited. People who would not support drug running, destructive chemical industries, apartheid, arms industries, forest destruction actually do all this through their banks.

Discuss banks and what they invest in and how socially responsible they are.

Types of economics

1. Informal economics

a. 1:1 direct exchange of goods with no money changing hands.
b. Informal work groups where people get together, each spending time at each other's houses doing such things as repairs and gardening.
c. Social work i.e. helping pensioners or disabled people in the community.

2. Formal macro-economic systems

a. As soon as exchanges are set down on paper they are subject to accounting procedures and this makes them formal economic systems.
b. These can be ethical or unethical. Ethical uses of money are socially responsible investment and community economic systems.

Socially Responsible Investment (SRI)

Ethical investment or socially responsible investment is actively diverting money from destructive to creative uses. Many people have some level of surplus income which is placed with organisations where it is used in socially responsible ways.

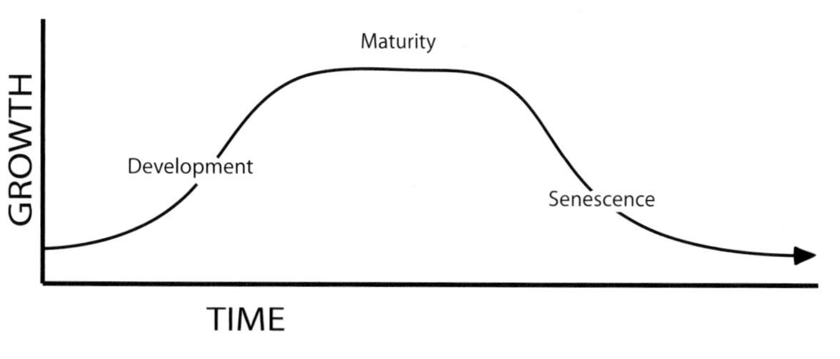

Investment in people and services

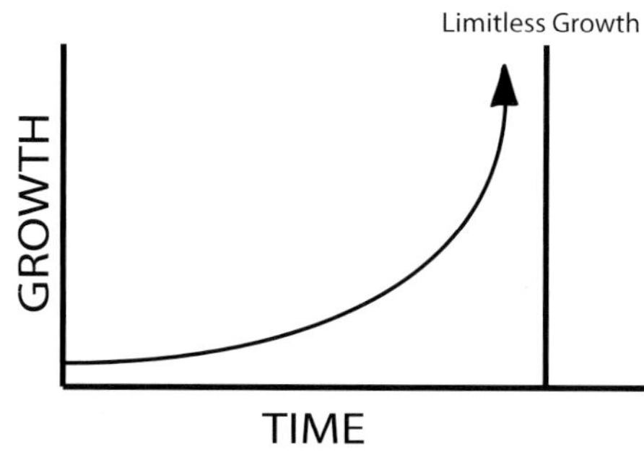

Investment in militarism

Examples of such organisations in Australia are:

- The Rudolf Steiner Organisation, e.g. Helios Pty Ltd
- Ethinvest
- Australian Ethical Investment Coy
- Maleny Credit Union

When investing money ethically there are three approaches:

- Divesting is the first level; identify what you do not want to invest in and taking your money out of them, e.g. Armaments, nuclear industry, Nigerian companies, also boycott their goods and services
- Affirming is the second level investment in things which are socially responsible e.g. Housing, jobs, forests, land rehabilitation
- A group identifies a company with good and bad products. They buy up shares and vote at a shareholder meeting to change the company policy in favour of good products. e.g. Greens bought into VW company in Germany and stopped it logging the Amazon

Community economic systems

These are economic systems which operate at the level of bioregions or smaller. Ideally, these economies are 60% of a total economy and the other 40% can be national. Some community economic systems are:

- Local Credit Unions
- Housing Trusts
- Community Revolving Loans Schemes
- Local Enterprise Trading Schemes (L.E.T.S.)
- Co-operatives

In Australia, the Earthbank Society collected and disseminated information about community economies, ethical investments, ways of using resources, consultants, and research.

Ask around the class – what do students really want to invest in and what they do not want to invest in.

Financial initiatives

- Permaculture Credit Union
- Slow Money
- Perma-occupy
- Barrett Ecological Services
  www.barrettecological.com
  'Freeing ourselves from money. Permaculture, governance and economics?' Interview with Leonard Barrett, President, Barrett Ecological Services, Portland, OR By Willi Paul, permaculturexchange.com

Two Case Studies in Socially responsible investment

Australian Ethical Investments

- 20% of funds must go to small businesses
- Save some funds always for trees
- Invest only in Australia
- Invest only in projects for a just and humane society
- Invest only in protection of the environment
- Will not invest in – uranium, woodchip, other mining
- Seven years operating and returns to investors were 19-24%

Bellingen Loan Fund

Analysis of the social structure of the Bellingen community and found:

- Rural economy
- High unemployment
- Often well-educated

Analysis of community money flows as follows:

1. IN
   - Exports – locally made
   - Government payments – pensions
   - Fixed Incomes – interest etc
2. OUT
   - Imports – purchases
   - Investments – banks etc
3. Aim of Loan Fund: Hold money in Bellingen
4. Need knowledge of the multiplier effect of money
   - No large developments
   - Analyse outgoings i.e. losses to the community
5. Investment Policy:
   - Local high quality goods
   - Human content services
   - Create demand for durable goods
6. Lending Policy:
   - Lend locally and buy locally
   - Reduce unemployment
   - Use TAFE for accounting.
7. Priorities
   - Emergency priority for compassion
   - Small business
   - Shared land
   - Last is consumer goods

Legally the Bellingen Loan Fund is a Company Limited by Guarantee which is non-profit and community based. Shares are $100 with a three month minimum withdrawal time. Interest on deposits competes with credit unions. Borrowers must be from the community. Investors can be from anywhere.

Student activities

Describe the type of socially responsible investment you would like to see in your bioregion? Describe in detail the activities and principles of such an organisation.

LETS System in Australia

There is a need for a supplementary money and trading systems in our communities – one that bring humans and ethics into trading goods and services. One that replaces unrestrained credit and growth. One that is a subsystem of ecology, not a threat to it. Real wealth or social wealth is when people co-operate within a community and working in a variety of meaningful and diverse activities. Real wealth is giving and helping.

Today thousands are stepping slowly away from the traditional economic monetary system and quickly toward the LETS system, which allow human needs to be realised.

LETS is an acronym for Local Exchange Trading System or Local Energy Transfer System. Some countries call it the Green Dollar system, though it is based on the original LETS that was developed in 1983 by Michael Linton, a Canadian.

LETS traders are groups of local people who agree to buy and sell within the group using a local currency as a means of exchange. The value of the goods and services is determined by those involved in the transactions, usually by the person who is offering the goods or services. Members are individuals or businesses. The management is run by a group of local people elected as a committee. They do not have to have any special skills.

Members of LETS register the goods and services which they have to offer and those they request through a notice board which is published with other information in a newsletter. Members receive the newsletters regularly. Members will browse through the newsletter for Offers and Requests and contact the person by phone. The value is determined and the transaction arranged. If materials have to be purchased as part of the transaction the amount payable could be part local currency and part federal currency.

When the transaction is complete, the LETS committee office is notified. This is done by phone or mail. The provider's account is credited and the consumers' debited. Each person receives a copy of their transaction.

It is important for members to realise that there is no debt in LETS. A negative balance indicates that LETS units have been issued to others, not borrowed from them. No one is waiting to be paid and no one has any claim to another's assets. No interest is paid. None is charged on negative balance accounts. If the transaction details of all members were plotted on a graph it would show a zero balance.

Most people new to LETS are concerned that there will be someone who spends a lot of local units and doesn't earn any local currency. Most systems set a ceiling level. If a member approaches that level the committee ask the person for an explanation and how they intend to start trading. Perhaps they are new to the area and intend to trade once they are settled. If the person has no intention of trading they can be prevented from trading. Because the system is local and neighbourly, it is self regulating – word gets around about undesirable people and nobody trades with them. Also, people tend not to leave the area in debit.

The organisation and methology of each local LETS is different. Although the principles of LETS are the same everywhere, local groups determine their own strategies and techniques of operation. For example, each system decides how often the newsletter is published, it is up to the committee to decide if they only print new requests and offers, or print the new and the old. Systems also use different methods to record transactions e.g. By phone, by cheque book.

Members of LETS find the system:

- Stimulates local interaction
- Keep wealth in the local area
- Empowers people with greater realisation of their self worth
- Provides access to goods and services formerly denied them because they are cash poor.

Given the above reasons, it is no wonder that the Western Australia State Governmment has awarded $50,000 for the establishment of LETS through the State Development branch.

And now the NSW government has awarded the Hunter region $47,000 for a similar project through the Local Enrichment Programme. The money is being administered through Newcastle LETS. Governments are realising the potential of LETS as a complimentary system to the federal one. However, it is paradoxical that federal dollars are necessary to start such a scheme. It is simply not necessary. Yet it is great that it will generate some with employment in the region and is an ethical use of federal money.

As far as the Tax Department is concerned, only transactions conducted in pursuit of one's usual livelihood need be declared. Due to the differences in value between local currency and federal dollars the person equates them as closely as they can.

There are lots of innovations arising from established systems, such as sub-groups in LETS which build stronger interpersonal links e.g. youthLETS, womensLETS. In Australia there has been an increase in trading between regional LETS (called interLETS). This enables a wider range of goods and services to be accessed by members which is particularly important when a system is starting and the range of offers isn't diverse. Some groups have individuals with federal currency, but no time to earn local currency, they can trade cash for local currency. On the other hand if you have an abundance of local units and are unable to spend them you can donate them to a community service fund administered by your local committee who distributes them to people with needs but not the capability to acquire local units or cash.

Bioregional Strength – A Snippet of History: India's Traditional 15th-century Taxation System

At a permaculture design course in Auroville Greenwork Resource Centre some years ago, Ardhendu Chatterjee, the centre's co-ordinator, explained how traditional India had a tradition of very rich bioregions. With Adhendu's permission I took some notes and summarise his talk here:

Imagine three neighbouring bioregions of rural India before the invasion of the British. Each had a central township or village and a clearly defined boundary.

Roads led out of the townships and connected up with townships in the next bioregion. Around the each central township and along the roads were taxposts.

At these posts everyone carrying goods had to pay tax when they left. There was no tax for people bringing goods in.

1. People were taxed according to the speed and efficiency of their transport and the weight of the goods.

   The lowest tax was for people on foot (say 5%), the rate went up for someone on a donkey (perhaps 10%), then a bullock cart (at 20%) and finally a fast horse carriage (up to 50%).
2. As they passed along their bioregional road, they paid taxes many times. The burden became heavy.
3. Ultimately, it became only worth exporting such things as gold, or very high priced luxury items.

Results: The resources were largely kept in the bioregion because it simply didn't pay to export them. All India with this system was locally prosperous. The system collapsed when the British wanted to control and export India's wealth back to England (the test case was precious timbers). They either made the system illegal, or simply destroyed the taxposts.

Imagine if in our countries and in our local areas the biggest trucks and the biggest ships and planes had to pay massive taxes whilst we, on foot and bicycle, could easily trade our home produce! What would happen to the world grain commodity exports? In fact, what would happen to the resource commodity markets and trade? It makes you smile just to think about it.

UNIT 33
Permaculture and ethical workplaces/ businesses

With more than 60% of the world's population living and working in cities many of us make our livelihoods through commercial enterprises (businesses) either as owners, directors or employees. So how to incorporate permaculture into business?

Unethical businesses and workplace practices are presently costing the Earth and many are in danger of meltdown as world economies teeter on collapse.

Offices, shops and factories are voracious consumers of non-renewable world resources, yet they have been largely neglected in permaculture design and courses. Many people think of permaculture as only about growing vegetables, yet permaculture principles have a major role to play in improving office environments and relationships. You can redesign your office or factory environment by setting up design aims and apply the ethics and principles. Refer back to Unit 3 and revise them.

Permaculture ethics have important contributions to make for business people as owners and employees. The concept of the triple bottom line in business, Environment, Economy and People applies our ethics perfectly.*

Ask and revise these and how they can impact on work places

- *Care of the Earth*
- *Care of People*
- *Distribute surplus to our needs*

Learning objectives

This unit is designed so you can use it for

- Planning and redesigning workplaces
- A checklist of where you are and how to move on positively
- A reminder of how good work can be

With tables and drawings used as posters you have easy ways to meet the design objectives.

* This unit was first developed by Margot Turner as Permaculture at Work and then added to by Lara Psychek as Permaculture in Business. Both permaculturists live in the Blue Mountains. This unit amalgamates their work

Ethics

Use the permaculture principles to design all operations in businesses and to create sustainable resilient, healthy workplaces.

Principles

The permaculture design principles for offices and industry are to:

- Reduce non-renewable energy, water and paper resources
- Replace non-renewable resources with renewable resources
- Support local people
- Reject or refuse bad products and workers
- Manage waste
- Establish a triple bottom line

Start small

- Work part time and build up
- Begin with minimal or no debt
- Leave room for mistakes
- Allow time and capacity for learning/honing skills/products
- Give yourself time to figure out if the venture is working

Do you enjoy it? Is it viable?

Diversify income streams

- To reduce risk if one has a slow period
- Remember that in business, as in gardens, diversity means resilience

Serve your community unlike national/multinational chains

- Co-operate more and compete less. This goes against the nature of business, because business is competition but the benefits are clearly seen from local patronage and support
- Help strengthen the community's economy and self reliance

Keep goods/services as local as possible (locally, state-wide) and work where it counts

- Assists in having financial independence
- Allow others to provide similar services to their communities
- Allow others to be in business too so co-operate with them
- Don't compete others out of existence to increase your market share. They are part of bioregional diversity

Search out and use local renewable resources

- Employ people instead of machines
- Develop less reliance on transport

- Boosts the local economy
- Enhance community

Do business with other environmentally responsible businesses

- Every ethical choice adds up to change and every bit helps (ask people to give their ideas)
- Develop new business ideas you can start working on part time

The triple bottom line is when an office, factory or company makes its decisions based on:

- Environment
- Social impact
- Economic outcomes

These three are considered equally and influence management decisions and when practiced, profitability and productivity will increase.

Design aims for work

The permaculture goal is to have its principles work for the three elements: people, property, and processes. It is monitored by regularly measuring the work ecological footprint.

People

- Value people for their diversity and provide them with a clean safe environment
- Support local people and suppliers
- Monitor and reduce their footprints

Property

- Design green buildings to use renewable, energy efficient resources
- Buy and use equipment which is people friendly, durable, safe
- Costs go down

Processes

- Work with low energy and renewable and local material processes
- Reduce waste
- Monitor the office footprint and reduce it
- Local custom increases

If you don't have design aims...

- All bills will be higher
- Light and air in the work environment will be unpleasant and perhaps unhealthy and staff may get sick
- Replacement costs for everything will escalate and come more often
- Will contribute seriously to climate change, and neglect the future
- Add to non-degradable junk placed in landfill
- The work place will be very vulnerable to major breakdowns and failures
- The energy and expertise of the staff may be wasted

Ecological functions for work

- Produce high quality goods and services
- Complement products and services with other businesses
- Provide social environment for pleasure and knowledge sharing
- Give value for time and effort
- Make work worthwhile.

Localising businesses

Apply permaculture design principles to your business.

Advertise your ethical practices.

- Start up and support local businesses that provide goods/services better than national or multinational chains. Ensure diversity in local businesses and create greater resilience and a healthier local economy. Local businesses employ local people and create relationships of trust to build cohesive communities
- Start up and support local markets as places to connect to the community for sharing, exchanging and selling goods and services
- As a business, get involved in the community and apply the same principles as those in Zone 0 through retrofitting workplaces or constructing new ones
- Purchase as locally as possible

Strategies for working better and enjoying it more while making a difference

Work by zones – work simply

Zone 1: The work kitchen and washroom

- Measure and reduce waste include energy and water
- Lunch in the nearest park with food made at home*
- Bring food containers e.g. Tiffin carriers for daily take-away foods and use crockery mugs instead of polystyrene
- Divert garden and food waste to a worm farm at work
- Reuse packaging such as boxes and paper bags
- Send out-of-date stock to community groups
- Monitor water and food origin and use

Zone II: The office environment: all that paper!

- Email or phone instead of fax and papermail

* Buying food and eating out costs the environment twice as much as food prepared at home.

- Handwrite replies at the bottom of letters and memos
- Keep mailing lists up to date to minimise wasted correspondence
- Work on screen
- Print drafts on used paper
- Design document with a clear purpose and audience
- Keep it simple
- Avoid using staples, plastic and wire binding
- Ask suppliers to take back packaging

Zone II: Reduce energy consumption: cost of office equipment is an energy and office equipment audit

- Purchase equipment with high energy efficiency ratings
- Use laptops instead of desktop computers
- Use ink-jet printers instead of laser printer
- Use electric kettles instead of urns
- Turn equipment off at the powerpoint when it is not being used
- Use photocopiers with low energy standby and double sided printing facility

Zone III: Buy where it counts: choosing suppliers

- Purchase local paper which is chlorine free and made from post consumer fibres
- Replace paper towels and tissues with reusable linen
- Use unbleached or oxygen bleached toilet paper

Property

Make buildings work naturally with low maintenance and toxicity, long lasting furniture and back up plans.

- Do regular energy and water audits. Identify where you can get the biggest savings for the least effort or cost. As a test, imagine that you have no electricity for three days, or no water. Can you still work?
- Use energy efficient technology and maximise solar energy
- Ask staff for a whole site water audit – including where the grey water is going. There are technologies now for cleaning office grey water
- Refer to clear water technology. Have taps which automatically shut off

Use environmentally friendly building materials and furnishings – refuse petrol based materials

- Use plant based paints
- Find vegetable dyed coir mattering
- Locate hemp fabrics for upholstery
- Use recycled or plantation timbers
- Buy chairs built to last which incorporate 'recycled' plastic or aluminium
- Purchase low emission fibreboards
- Use steel rather than aluminium products

Emergency and back-up elements

- Design for water tanks or ponds
- Employ sheltered places to work outside
- Design an emergency energy plan such as: work at the library or other public places, and have an alternative supply of energy such as gas if the work place is electric

Care for people: work and feel well practices.

- Treat each other with respect and dignity in the workplace
- Create safe and healthy working environments e.g. Have meetings in parks for the benefit of fresh air and a little exercise walking to get there
- Share skills and knowledge

Workplace lighting

- Arrange workstations to make use of natural lighting
- Use desk lights which use a fraction of the power of standard mounted ceiling lights
- Work in common areas of work and use less light than individual work areas
- Zone lights so that they can be turned off when not in use
- Clean windows
- Switch off lights when not required

Minimise the impact of volatile organic compounds(VOCs)

- Ventilate – let more air in – open windows
- Introduce indoor plants to absorb toxic indoor chemicals
- Avoid air fresheners
- Avoid chipboards as they give off formaldehyde vapours
- Small offices use public photocopiers from libraries and post offices

Heating and cooling

- In winter use sun on glass to warm buildings
- Ventilate and allow cross-breezes in summer
- Dress for the weather not for the air-conditioner
- Send toxic materials to your nearest hazardous waste facility

Transport

Walk, cycle to work or car share and have car free days and catch public transport.

Work together

Treat each other well:

- Remember work is something of value to yourself and others
- Change behaviours that aren't working
- Think about the global and local impact of your work
- Find ways to work with diversity.
- Talk about the value of work to yourself and others
- Treat each other well
- Find ways to change behaviours that aren't working
- Monitor how your work is affecting the environment

The global and local impact of your work: ways you can work with diversity

Support local people

- Buy locally produced ingredients and materials preferably organic
- Make sure to ask your supplier for their environmentally responsible purchasing program
- Become involved with local environmental organizations and community events
- Support a local landcare group

Retrofitting your workplace

Reduce consumption:

- Make micro changes at work and encourage others to do the same, like car- pooling, recycling, choosing not to print emails, growing lettuces in containers by the window for fresh greens, reducing packaging, making your own lunch because take-away food uses far more resources. Eventually small contributions can become work place movements or at least make big differences
- As an owner/director/manager, structure or re-structure business towards sustainable business practices to reduce your ecological footprint. Do an audit of water, energy, transport, furniture, materials, paper, etc. and work to reduce the consumption of these resources by making quality products
- Make environmentally sound changes – this will buffer your business for times of resource shortages and rising costs
- As an employee choose to work in businesses that have or are working towards sustainable business practices and assist by making suggestions
- As an owner/director/manager, make high quality, durable products which do not need replacing
- Do business with other environmentally responsible businesses offering quality services/products. Make an SRI inventory and maintain adherence to the triple bottom line
- Constantly check your ecological footprint of the business and each employee

In conclusion, this unit is based on the concept of Right Livelihood which does no damage to people or the environment.

Student activities

- *Carry out energy, water and waste audits at work and discuss the results with co-workers. Make a priority list to reduce it*
- *Do your work ecological footprint now and in one year's time*
- *What is the most likely disaster for the place where you work – see Disaster Unit*
- *Make a plan to endure it*
- *Redesign your office, shop or factory so it works better*
- *Think about each of the five zones*
- *Zone I – the Kitchen and wet areas*
- *The water collected on-site, energy collected and stored on site for power, heating, cooling ...*
- *Can this built environment be changed to work better?*
- *How much of the food consumed is grown on-site or locally? For teas, salad vegetables ...*
- *What is the take-away container count? Can you get it down to zero?*

UNIT 34
Land ownership

Land ownership, like money, is an emotional topic since land is most commonly used as a commodity for making profit and not a resource for meeting human needs. Who holds the Title, and how it is recognised can have a big impact on human relationships. This Invisible Structure can make and break communities.

Many people want to share land and live on it in communities. However their hopes can be destroyed by simple lack of knowledge about ownership.

Land ownership differs in different places. Laws about land ownership change quite frequently. The notes here are a guide only and teachers will need to adapt them. They are not a legal document. This unit is about permaculture views of ownership.

Land ownership must one day finish. Every person has a natural right of tenure to a piece of land to supply their own needs for food and shelter. Today the cost of land does not reflect what the land will produce. The unrealistic price of land is one reason that many people purchase land together. The realistic thing to do is to lease a 'right of use'.

Approximately 50% of all Australian forests are privately owned. In the USA, conservation purchase trusts have been set up which buy large areas of wetlands to take them out of exploitative hands. Ownership can produce ethical land-use.

When working as a permaculture designer always ask about what sort of land title the site is under and how it works.

Learning objectives

In this unit, students will learn to

- Differentiate between land rights and ownership
- Describe several types of title
- Select a good title for communities
- Describe individual rights to land

Ethics

Ownership has responsibilities and land must always be left better off than when the user acquired it. In group ownership a good ratio is to have 20% of land held individually and 80% in common with clusters of 3-12 houses. Always ensure that the title is not held away from the design site.

Principles

- There is a title suitable for each situation and purpose of land ownership. Wrong title can destroy human relationships and lead to land exploitation
- Permaculture designs with their resources have some times been implemented on land which was later arrogated to private use
- Designs and implementation on public land must be guaranteed for maintenance and future public access

Ask students what they know about or what their experience is, of land ownership. Draw on this throughout the unit. Keep discussion open.

Titles

Napoleonic code
Strata title
Torrens title
Freehold
Community title
Customary title

Land ownership titles

- Individual ownership – One person holds the title and the deeds
- Group ownership – In NSW, Section 40 which governs the density of housing in rural areas, applies to mean that it is possible to have only one dwelling per 100 acres. Councils differ about this and an objecting council can be difficult. If buying into a community, make sure it is an approved Multiple Occupancy or Strata Title
- Land can be owned by an Association Inc, a CLG (Company Limited by Guarantee) or CLS (Company Limited by share), a Pte Coy (Private Company), a Co-operative or a Unit Trust

Tenants-in-Common With this system there is no defined ownership. More than one person owns the land. If the land is owned by six people, each person owns 1/6th of everybody's house, crops etc. If a person wants to leave they can demand to be compensated to 1/6th of the value of everything. With a partnership of two shareholders this works well. An agreement is drawn up as to the conditions of ownership so that if one wants to sell the other gets first option. The new buyer purchasing half is bound by the original conditions in the agreement.

However there has been a change to this law in Australia. Ownership is now defined under

Tenants-in-Common legislation by drawing up a boundary which is defined in terms of percentage of things on the land and physical area. Check the law in your area.

Multiple occupancy (MO) It has ill-defined ownership and has caused many communities to fail. Generally banks will not give them a mortgage. However it can work. An example is Penrose Community in NSW. Until 1990 Multiple Occupancy was the only title applied to communities in NSW. A code of the rights of individuals in MO has since been passed. The details can be obtained from NSW Department of Planning.

Since it is not possible to have individual title, strata title is excluded. The individual has an agreement with the incorporated entity – possibly a lease giving each person exclusive use of a portion of land. The incorporated entity holds the title and grants leases. Individual home areas are from 0.5-2.2ha. A First Home Owners Grant is possible in Australia.

Unit Trust In a unit trust individual shareholders own so many units of land. Crystal Waters, Queensland, is set up along these lines. The houses are in clusters since it is easier and more economical to supply energy and water, and land disturbance is minimised. One hectare is allocated for individual gardens (0.5 ha is more than sufficient for intensive gardening). Homes can be private. Everyone can have a good aspect and use common facilities. Buying and selling is relatively easy.

Strata Title for Community Rural Land
This gives individual rights to land. Any improvements are covered if people want to sell. In NSW it is called Community Titles Legislation, and in Victoria it is the Cluster Title and in Queensland it is Group Title.

The Land Titles Office in NSW has information about the new Community Titles Legislation. Each council can make its own choice about how to apply it. Always ask for council provisions for housing density on rural land if you dream of one day living in a community.

Customary title Occurs where people live on communal lands such as in Botswana, and other countries of Africa. People in clans can share land for traditional purposes such as grazing. It can also be thought of as Aboriginal title.

Napoleonic title Exists where countries had French colonial governments. The land is inherited equally by all members of the family. Finally the parcels of land are tiny and separated over the landscape. Now it is common to try to get families and individuals to regroup their lands to work in more efficient units.

The Vietnamese were allocated land according to their needs by the local People's Commune Committee. This was based on need for land to grow rice. However as family size has changed and farmers migrated to cities, the land is not distributed equitably or well for agriculture.

Student activities

- *Research title for individual and groups in their areas*
- *Report on what density of housing is allowed on local rural land*
- *Determine what sort of title is best for intentional communities and why*

UNIT 35
Land access and ethics

This unit is a set of guidelines for acquiring access to land and taking ethical responsibility for it. Having access to land can be considered a natural right if it is for meeting personal needs such as food, shelter and fuels. To leave land better off than when it was acquired is a natural ethical responsibility. This unit discusses several ways to gain access to land.

Land is easily abused or destroyed when it is seen only as a cash reserve to be bought and sold for profit, or when its resources of minerals, trees etc. are seen only as cash. Land is also abused when people live far away from the impact of their lives on it. For example, city people continue to consume huge quantities of resources from rural areas without knowing what pressure they put on the land. Nor do they realise that their lifestyle is actually damaging the Earth. Yet the number of landless people in the world is growing. In cities and rural towns, people need access to land to supply their own needs.

This unit is not a legal document and each person must take responsibility to research the laws about this topic and ensure they are compliant.

Learning objectives

By the end of this unit, students will be able to:

- Establish a set of ethics for care of land
- Know how to get access to land to grow food
- Describe three land access strategies

Ethics

Permaculture holds the ethic that every person has the right to land for shelter and to grow their own food. However there are ethical responsibilities to holding and using land.

For single ownership individuals write an ethic for their land.

For community (Bioregional) landholders:

- In bioregions, ethics are written into a document. Possible examples of such ethics are:
- Wise use of resources
- Encourage sustainable lifestyles
- Create life balances etc

These ethics are also written into the charter of each organisation having access to that land:

- The Land Trust or Co-operative which holds the land
- The Credit Union, Trust, Co-op or CLG (Ethical Company Limited by Guarantee) which carries out the financial management
- The Co-op, CLG which manages the enterprises
- The Equity Trust or Trading Trust which carries out the business/trading function

Then each of these has a business ethic.

Terms

Ethical land use Preserves and increases biological wealth such as soil, water, trees, species.

Land access Every person has the right to access to land as a resource for food and housing. Land should not be treated as a commodity.

Each person (or group) write their own ethics for land in their care. Ask students to read them out and discuss them with others.

Principles

- Land is best held as a Trust, not individual ownership
- Do not export anything from the land that will reduce its sustainability
- Water is a priority in selecting land
- Surplus energy can be a product

Land access strategies

Informally land access

Approach individuals, local government bodies such as schools, churches, neighbourhood centres, and so on.

Formal land access for groups

A Land Trust buys units of land and peripheral town land. It also buys villages as food clusters and puts covenants on the land. Councils are urged to donate to its Green Zones. The land can then be given/allocated but the trusteeship and its goals are still retained. This way there is no taxes, no inheriting relatives, no capital concerns.

Land Planning and Land Use

- Usually planners who make the decisions for people's lives are brought in from outside. Tertiary training often creates local resistance
- Set up fast communication systems with little bureaucracy and use local people from the bioregion
- Do not export anything from the land that will reduce its sustainability e.g. No uranium or no aluminium smelters. Resources need to be

categorised and recycled

- Ideal exports are seeds (surplus), information, music and crafts
- The very last thing to be exported is grain crops. If they can not be grown locally then try to find something else to eat. They are very exploitative of the land. They mine the soil badly
- Question dietary habits because most of the worlds grain goes to cattle
- Water needs priority. River catchments must be protected. Efficient and environmentally safe usage of water must be encouraged
- Surplus energy can be a product e.g. Wind or solar farms or hydropower

Land rights

Permaculture holds that it is a fundamental right to have access to land to supply basic needs such as shelter and land to grow food. It has nothing to do with ownership. Land is not a commodity to be bought and sold. It is an essential life-sustaining resource.

- Become individually responsible for any land in your care
- Put land in trust so it is cared for in a sustainable way so that anything exported from the land will not deplete the local resources
- If it can never be bought and sold it cannot become a market commodity

Public access strategies

Set up a land access office which is tailored for the use and potential of each bioregion in order to develop access to land for people who need it. Following is a list of several tried land access schemes.

Oxfam model

This is a land-lease system. It has a regional office which makes up two lists and puts them in local centres like libraries, health food shops, CES.

List A. For those who want land to grow food:

- In high-rise situations and other crowded areas
- People with special needs such as young families
- Those in rental housing

List B. For those who have land and are willing for people to use it:

- Elderly people
- Those too busy
- Those with large gardens
- Can include public land owned by Waterboard, Council etc

The regional office works out an annual renewable lease agreement between two parties. Payment arrangements can be made either with money, a percentage of the produce or by doing chores. The central office makes a small service charge for negotiating leases. It is important that there is a third party involved in the drawing up of a lease.

City farms

- Interested groups negotiate a lease with their council never for less than five years and with an opportunity to renew for perpetuity. These can vary in size from 1-100 acres. Many are in industrial areas
- The city farm can be a CLG and must register with the Corporate Affairs Commission or equivalant
- A management committee is appointed and within two years from the start it should be running on payment (entry fees) from visitors and schools. If it is not fully independent within five years something is very wrong
- It can include: nursery, worm farm, tool rental, recycling centre (including furniture), demonstrations, allotments and community gardens, domestic animals for farms, family-community meetings and picnics, gleaning, seeds and books, resale items, sales and a seminar centre

The city as a farm

This is when an individual or group decides to harvest one or several city products e.g. mature timber for furniture and mulch, chestnuts, grass clippings.

Farm link

- This is a producer-consumer co-operative. Twenty to fifty families in the city/town link with an individual property say, within one-and-a-half hours' drive of the city. The families involved spend a certain amount of hours (e.g. four days per year) on the farm helping with harvesting, mulching, composting etc. A food distribution network is organised
- Quarterly meetings are used to work out exactly what products the consumers need and the farmers grow them
- The Co-op may also negotiate farm holidays and educational workshops

Commonworks

This is a farm held by a land trust and is close to a town/city. It arranges a whole series of special leases on the land for such purposes as forestry, livestock, crafts, teaching space, cut flower production, nursery functions and mudbrick workshops. 10% of the net income is paid back into the Commonwork Fund and this pays rates and maintains the land. One example is Commonwork, Bore Place, Chiddingstone, Edenbridge, Kent, TN8 7AR, UK.

Farm and garden clubs

A farm is purchased, usually on a public access route from a town or city. People buy shares. A manager is appointed. Sometimes it is a good way for farmers to get out of debt. Motel-type accommodation is developed and special interest enterprises such as aquaculture, retreats use, health, nutritional and healing natural practices are established.

It is absolutely essential to have firm legal agreements on access and structures. It needs a 'lean' management set-up with, say, two to four people involved in planning and consulting with the rest.

Arrangements are made on friendship, ethical, social values and with minimum membership fees. It is enriched if the people involved can offer other services such as legal help, accounting, Permaculture design and so on.

Tourism as land access

- Tourism has many negative aspects. One is the ever-widening gap between the 'have's and the 'have-not's
- It is often a big strain on local resources. Towns cannot cope with water demands placed on them. Sometimes local people suffer because retail prices are raised for tourists
- In non-capitalist places results can be worse with local impoverishment, cultural crises, youth dissatisfaction, and a zoo-like feeling by locals, e.g. Alice Springs and Mauritius. Locals become modern 'slaves' to rich tourists who have no commitment to the bioregion

Positive tourism, if it exists at all, must enhance the local economy and preserve local resources. Bed and breakfasts can be OK and also local environmental study camps.

Student activities

- *List what they think could be desirable tourist activities in their region*
- *Design a system that gives every person in their bioregion access to land to grow food*
- *Write the ethics for your Bioregion*

UNIT 36
Legal structures

Legal structures are the types of society or company that ethical communities, land owners and local organisations get involved with. The wrong one for the wrong purpose can cost time and money and such invisible structures can create problems and insecurity for people.

Most community groups have to do some sort of business and/or handle money. Some examples would be a food co-op, day-care facility, or a company to do business e.g. Teaching, printing. Many need to form formal organisations and are often concerned about the type structure they need to work under.

This unit is more important than most students realise. Many feel it is very boring if they have had no need for this knowledge. However, those who have spent time getting 'incorporated' in a community organisation, or tried to set up a private company, find it really interesting. It is relevant to students who might join communities, start local co-operatives, or become involved in activist work. It is fairly easy to understand and students like to talk about their experiences. You need to update or localise your knowledge.

Learning objectives

By the end of this unit students will be able to:

- Describe the strengths and weakness of four structures
- Say what structure would be best for a group they know
- Explain why the NSW Aboriginal Land Rights Act is weak

Ask who has ever been in a position to create a legal structure as a committee member or otherwise. What was their experience?

Ethics

Structures are ethical when they lead to resource renewal and serve the right livelihood principles of its members.

Principles

There are structures that suit some organisations better than others.

Discuss the Aboriginal Landrights Act and its legal weaknesses. See end of this unit.

Types of legal structures

1. Incorporated Association

This is a helpful structure and does not need a solicitor. The name and charter is registered with the Department of Corporate Affairs. Incorporation costs $78. If there is land, it can be owned directly, or held in Trust. e.g. Community Title could be held by the Incorporated Association.

2. Company Limited by Share (CLS)

This structure is a bit awkward. It is limited to 50 shareholders, and is mainly used by people involved in multiple occupancy titles.

3. Company of Association Limited by Guarantee (CLG)

This is a non-profit making company. A CLG is a good community vehicle and carries its members' philosophy e.g. Anti-racist/anti-sexist. It has members, not shareholders, and they have no right to income but the employees do. Members guarantee a certain minimal amount if the company goes bankrupt (usually about $50).

There is no automatic limit to the number of members, however, a limit can be imposed if desired. It is possible to have different types of membership, e.g. Permaculture International has life members, non-voting and so on.

The profits are paid out to the company according to its objectives and not to shareholders. If the company is wound down the profits go to another organisation with similar objectives. A business can also be a CLG and trade between members is tax exempt. It is cheap and simple and generally has no tax problems.

There is a nearly identical CLG structure in all Australian states.

4. Proprietary Coy

A private company can be a good tool for a non-profit organisation if it doesn't need more than 50 shareholders.

It has up to 50 shareholders, costs $800 and can be bought within 24 hours. It is a capitalist vehicle. Simply phone 'instant companies'.

5. Co-operatives

Use of the word Co-operative is illegal unless the business is incorporated. Co-operatives have shareholders and members. Both have the right to income and to own shares (capital of the company). Co-ops exist for:

- Workers (must pay award from day 1)
- Rural affairs (communities)
- Trading
- City communities

Co-operatives are used for community organisations. They have very rigorous auditing procedures and it is difficult to change the rules. Some say they contain archaic, inflexible legislation. (The Co-operatives division of the NSW Dept. of Consumer Affairs has recently given grants to stimulate their growth.)

6. Trusts

A Trust is simply a document lodged with the Registrar of Business. Trusts must have beneficiaries. The Trust document should state the aims, objectives, and responsibilities of Trustees and how to handle whatever is being held in Trust. Use a corporate business or association e.g. CLG, rather than an individual as Trustee. The laws are constantly changing. Trusts exist for:

- Property
- Equity
- Charity
- Discretionary (for families)
- Unit (corporate and property)

Trusts are good to gift and receive land, to trade leases and for rights of use.

Try to separate assets (capital) from trading activities (income) – have no possessions (capital) but have access to them. Trusts can have various business names, Tagari Publishing, Phoenix Seeds, Earthworks Trust...

For example: A Company Trust can hold all the land, fixed assets, machines, vehicles etc. Another Company Trust can trade and earn income.

A unit trust is the most common trust for owning land where individual shareholders own so many units. Crystal Waters is set up along these lines. A unit trust owns the land whereas individual shareholders own units.

Bill Mollison recommends all land be owned as a trust and its use limited by a charter. For example, create a land trust of your land and let your inheritors use it if they support, abide by and agree to the charter.

Student activities

- *List one structure which you would set up for holding land*
- *Describe how it would work*

- *Describe the best sort of structure for a community organisation which you are a member*

CASE STUDY

The 1983 Aboriginal Land Rights Act in NSW: A Legal Structure which cannot give Security

This Act, which created the Land Councils and gave them title over Tribal Land and Land Claims, is not 'enshrined' in the Constitution. This means that the right of Aboriginal people to hold Title to land was able to be set aside by any subsequent government which can change this law. Sunset clauses do the same thing by allowing the law to lapse at a determined time.

Non-Aboriginal people buying land and obtaining Title can only have theirs revoked by essential services and then they must be compensated to the full market value. Non-Aboriginal Title is secure. Aboriginal title is not. Hence Aboriginal people would gain greater security through acquiring land under ordinary (non- Aboriginal) Title which is not easily set aside.

UNIT 37
Designing communities

Today, many people want to live in communities and, for those who want to care for land, communities can use resources efficiently and minimally. In some countries governments resettle people to ease pressure on the land or cities, or because people are refugees. All these types of resettlement communities will continue.

In general people are not socially or economically prepared for living in community. Often the results are negative because people lack land and management planning strategies. In developing countries new settlers survive by cutting whatever forest remains to sell the wood to buy food. While wealthier settlers in consumer countries fill in their days sitting on and pushing mowers.

There are only two resources: land and people. Either of these can make or break communities.

This unit explore success and failure in intentional communities whose objectives are for reducing consumption, sharing resources and living peaceably with neighbours.

Learning objectives

In this unit students will learn to:

- Explain why some communities fail and others succeed
- Describe a community that has succeeded
- Use some broad design principles for communities
- Explore ideas for community management

Teaching tools

- By now you, the teacher, are introducing, integrating and revising all the past work
- Students are actively participating in class discussions and ideas
- Make you own graphics and re-use those from earlier lessons
- This unit provides useful information for students' final designs
- A sandbox, a model or feltboard with elements are the most useful tools for this unit

Ethics

The land must always become more sustainable, in species diversity, soil improvement, water and conservational usage. Management systems give maximum freedom for responsible development of human potential.

Principles

- Pay attention to and work with only two resources: land and people
- Communities provide good opportunities to use resources sparingly and replenish common land with water, biodiversity and soil

Problems of new settlements

Ask students for their ideas of what the problems might occur? Group the answers according to Land, People and Title.

Land

- Cheap, marginal land
- No capital for roads, dams, fences etc
- Isolated and difficult or costly to farm

People

- Lack of realistic objectives – too idealistic
- No forward planning – annual, monthly contributions
- No framework for decision making
- No design for orchards, water etc

Title – Ownership

- Ill-defined ownership
- Banks won't lend to Multiple Occupancy
- Legal status dubious – no voice in local government because of illegality
- People leaving can break communities
- Until 1990, no appropriate title in NSW

Success factors in communities

Ask students what they consider the success factors are. Again list under Land, People, Title and Other Factors.

Land

- There must be a full discussion and total agreement of the ethics of the community. e.g. Protect water supplies
- Obtain agreement for an outside designer to make a land management plan then community members add necessary changes – then stick to it. It is better for the whole community to do a PDC and agree to their own design
- Establish limits to what numbers an area can support and concentrate on optimum numbers. Stocking rates are critical for people and animals
- Permaculture villages can become tourist attractions. Some visitors want to work there for a while. Establish visitor criteria

People

- Only hold meetings for work
- Trust people in their area of expertise. Trust often reflects people's own level of responsibility
- Levels of public and personal privacy must be defined and clarified. People must be aware that there are private areas in villages

Title – Ownership

- Local residential ownership must be a minimum of 60% and higher if desirable
- Appropriate Title is necessary
- The time for leased and long term rented land must be at least 20 years

Types of community management

Discuss with students what types of management they know of. Do they have any ideals?

Authoritarian

One person rules everyone e.g. Religious/army. It works for a while if that person is competent and benevolent. Eventually, there are splinter break-away groups .

Majority rule

Usually divide-and-rule. This ignores the wishes of minority groups who can be very important and useful to society. Then minority groups either won't co-operate or they leave and form another group.

Anarchy

No ruler, no rules, no ruled. Differences of attitude can result in alienation and confrontation. Theoretically, anarchy is the highest level of responsibility to the personal and collective good. However, it seems that people are not yet perfect enough to be able to accept this responsibility.

Consensus

No ruler no ruled. Equal voice and right to opinion and usually no criteria for opposition. It can give power to one individual to hold up decisions by withholding their agreement. The power to boycott can result in stagnation over a long period of time. Quakers have a sensitive and workable 400 year old tradition of 'consensus', which requires everyone to abide by a clear predetermined procedure. Without this process the decision making would fail.

Appropriate consensus decisions for communities

Mollison says that there are two occasions when consensus is necessary:

1. When everyone accepts the ethics, design, aims/ objectives, methods.
2. Everyone agrees to have no more consensus decisions.

Trust is necessary from then on.

Permaculture management by hierarchy of function

In communities and society, how people function is more important than their status.

Explore this statement with students. Think of examples. The idea was first introduced in Patterns.

Individuals or small groups of individuals take responsibility to do a job and to do it well. Expertise and ability is respected. Hierarchies of function are then determined for the following:

- Water Supply
- Bulk fuel
- Forestry
- Road maintenance
- Crèche
- Financial management etc

Meetings

Keep meetings to a minimum – don't have meetings for the sake of them.

Have meetings only for work and ask how and when, not who and why. The who and why get discussed in social dialogue (over the fence) rather than in structured meeting. Otherwise the community will get endlessly bogged down.

E.g. The road was not maintained. So ask how will we fix it and when? – not why it wasn't fixed and who did not do their work. Whoever did not do their work knows and can be left out next time or they will do their work. Discussions about why and who are often speculative and destructive.

Case study 1

Village Homes, Davis, California

Land
20% of land is held privately and 80% in common.

Energy
Houses in clusters and every house placed for maximum solar gain. There is actually a caveat on each piece of land so that every house must come up to a minimum requirement for solar design. Every home pays 40% less than a conventional township for energy.

Green Belts
The front of each house faces away from the street and is focussed onto gardens or green strips of orchards, corridors and cycleways. Much food is grown along the corridors for commercial enterprises such as food and flowers.

Clusters
Hamlets of houses within the village help people retain a local community feeling.

Traffic
Circulation patterns for car, foot and bicycle are designed to encourage foot traffic and to protect it. Community gardens are placed where children play so families can be together.

Production and recreation are not separated. One aim of the design is to encourage productive recreation.

Tight ecological controls
Creeks and waterways, roads and ridges are in forests.

Social patterns were looked at to integrate ages and interests.

Case study 2

The Quaker Settlement, Whanganui, Atearoa New Zealand

Ethical communities

Invisible structures should permit maximum individual freedom with hierarchies of function rather than conventional decision making.

Population
To make a village really work there should be from 100-3,000 people to support enough livings to integrate a society e.g. School, dentist, healer, baker etc.

Ecological controls

Right at the beginning set up Environment Protection Goals for:

- Soil
- Cleanse
- Build
- Water
- Rain
- River
- Animals
- Wild
- Domestic
- Plants
- Indigenous
- Introduced

Energy must be renewable.

- Solar based on aspect
- Wind and water on gravity
- Biogas from compost etc
- Plants – for biomass etc

Design for villages and communities

Design with the same Zone and Sector Analysis as for a farm, using overlay techniques.

- Zone I Homes and food first(security) gardens
- Zone II Close public space and orchards
- Zone III Larger open spaces and community gardens
- Zone IV Reserves, fuel forests, windbreaks etc
- Zone V Wildlife corridors, native plant and animal sanctuaries

Student activities

- *Work in groups to design an ecovillage or, eco-neighbourhood of a suburb or city*
- *Decide on the type of management*
- *Write ethics for a new settlement, or a retrofitted one*

UNIT 38
Transforming the suburbs

You now want students to be able to analyse rural towns or suburban neighbourhoods then redesign it. This unit is thrown over to them. A few years ago a DVD circulated demonstrating that suburbs were deserts. However this has been rethought. Suburbs have many strengths and resources which are underexploited or discarded. They are now seen as being obvious places for developing resilience and sustainability .

Teaching tools

- Visit the neighbourhood, or quartier or rural township, and ask students to redesign it during this lesson
- Take big sheets of paper and coloured pens
- List the criteria for sustainable suburbs

Ethics

Suburbs need to 'stand alone' and be as self-sufficient as possible especially in water, energy and food.

Principles

It is unusual for permaculturists to design new suburbs however increasingly involved they are in retrofitting them.

Sustainable suburbs meet the following criteria:

- Local transport favouring bicycles and foot traffic
- Dispersed food gardens and orchards which meet tall local needs
- Local stand alone water supplies, energy and food
- Local markets and small clean businesses
- Calmed streets
- Long term harvest timber and non-timber tree forests

Modern suburbs are deserts producing very few of the goods and services that they consume. However there are many new movements such as Sustainable Streets, Verge gardens, and eco-neighbourhoods.

Ask class to work in groups on the following questions for half the lesson and then give their results for the rest of the lesson.

- *What are the characteristics of suburbs?*
- *How can they produce more goods and services?*
- *What is your idea of good suburban design? The results should cover some of the following points:*

Strategies for suburban revival

Co-operate by removing fences and do permaculture designs for a group of homes to improve facilities and amenities, and do permablitzes for individuals.

Encourage small and clean industries such as 1/8 ha of fruit, nuts, honey, poultry, and an animal lawn mower.

Harvest resources such as rainwater by using tanks, and purify greywater. Harvest old trees for timber and fuel. Store surpluses such as apples, pears, onions etc.

Harness human potential, use clubs, worknets and get self-responsiblity accepted. Use groups to plan streets, and identify community needs and wants e.g. Laundries and childcare and fulfil them.

Diversify the suburbs e.g. Permaculture Suburban renewal. Convert carparks, school, railways lines, etc. to productive parklands.

Techniques

Food

- Grow all their own food, flowers and fruit trees
- Start community gardens for rental housing people
- Use front gardens for food and climate control
- Use trellises for all vertical plantings
- Include special restaurants and supply special foods e.g. Wild rice
- Create jungles

Animals

- No dogs, cats, pools, mowers, lawns, or 6ft fences
- Use animal mowers for active open space
- Encourage indigenous plants, insects, birds
- Meadow lawns of herbs and flowers with weeder geese.

Energy

- All energy systems hydro, solar and wind renewable
- Set up examples of all the above and sell and spread information
- Total waste recycling

People

- Promote co-op housing and group activities
- For anarchists there is green guerilla work
- Markets for artists and craftspeople to meet local needs

Answer the 20 Questions in the Where You Live Survey

Case study

20 Questions about your neighbourhood

The Shape of Where You Live

1. What feels like the centre of the neighbourhood?
2. What are the boundaries?
3. Where is the high ground/important place?
4. Where is the low ground/least noticed place?
5. Where is the water. Is it still or flowing?

Notice How Your Place Celebrates Sacred Time

1. What and when are your festivals? (Be sure to include local festivals like garden visits, and garage sale times)
2. Where do people go when they go around the neighbourhood?
3. What are the holiday/ritual objects e.g. Fireworks, public art?

Find Out What Lives In Between

1. What is in the alleys?
2. Where can you go to see wild birds?
3. Where can you go to get away from cars?
4. Where do adults go to get away from children?

Find Out Where the Spirits Live

1. Who lived here before you?
2. And before them?
3. Where have the scandals and tragedies occurred?
4. Where is the haunted house?
5. What happens in the neighbourhood between midnight and 5am?

How Healthy is the Soul of the Neighbourhood?

1. What grows in the neighbourhood e.g. Trees, flowers, weeds, nothing ?
2. What places do children play and children's names for places?
3. What makes your neighbourhood different from the next?
4. Where can you go to talk to someone who is not a member of your family?
5. What basic needs can you satisfy in your neighbourhood by walking or bicycle?

Student activities

Redesign your suburb or neighbourhood

Place a large sheet of paper on a big table and mark in typical neighbourhood buildings. Students gather around and with coloured pens, retrofit the design. List the criteria:

- *Renewable energies*
- *Bicycle friendly*
- *Grow own vegetables in each garden*
- *Close fruit trees and orchards*
- *Community forests*
- *Community meeting place*
- *Footpaths*
- *Autonomous in water*
- *Increase protein sources.*

To do this:

- *Eradicate all lawns*
- *Calm the traffic*
- *Add to biodiversity with wildlife corridors*
- *Surface water*
- *Create no waste*

UNIT 39
Designing cities

Cities are increasingly congested and polluted. Permaculture offers some solutions to these problems to create greener and low carbon city environments.

The old idea was that urban renewal would kill cities however with the new principles we can restore life to cities. We must remain unimpressed by huge development.

There is a lively and growing eco-city movement. The conferences are vital and challenging. Some cities are taking the lead in recreating themselves. Seattle is one of these. China has now told its designers that for every new city there must be space to grow the food for the people who live in the city. As carbon consciousness grows, cities are setting goals to reduce their carbon footprint and associated with this are cleaner forms of transport, cycle paths and localisation as in the transitions towns movement and the bioregionalism.

Every city is recognised to a greater or lesser extent by the following:

- An unnatural ratio of hard:soft textures
- Wind patterns that are dramatically altered and called 'canyon effects'
- Temperatures are elevated by radiated heat from bitumen, plate glass and walls of buildings
- Increased run-off from rainfall and pollution of water and air
- Creeks and rivers have high CO_2 levels
- Lack of plant material
- Many rodent/scavenging type animals e.g. Rats, cockroaches, seagulls and pigeons
- Very high pollution levels and a deficiency of negative ions
- Great dependence on fossil fuels, motor vehicles and electricity. (In Sydney, 20,000 tonnes of pollution per day is pumped into the air from cars.)
- A high dependence on chemical, imported water. (New York city can no longer carry out the level of repairs required for water mains. More water is lost than is used.)
- There are many empty buildings due to high rents or dereliction

Cities have low levels of self-sufficiency particularly for food. They are voracious consumers of resources and cannot look after themselves. Social dislocation is pronounced, and a fuel or food shortage is an emergency. Hence repressive legislation is passed so the city will not be held to ransom. This leads to a reduction in human freedom and choices e.g. Right to strike of truck drivers.

A city is an IMPORT MONSTER with bureaucratic systems vulnerable to strikes and shortages. But it is not all bad. One of the big permaculture challenges is to green cities and to increase sources of clean food, energy and water.

Learning objectives

At the end of this unit students will be able to:

- Describe design principles to retrofit cities
- Design people-friendly areas
- Select some strategies for immediate use

Terms

Phytomass Total plant materials – all plants, roots, leaves, flowers etc.

Pseudomonas A disease causing organism.

Ethics

Cities need to produce more renewable materials than they use.

Principles

In the last few years, two sets of principles for cites have emerged. One is for new cities and the other for retrofitting cities.

Principles for the ecological city

1. Minimum intrusion into the natural state
2. Maximum variety of land use and activities
3. As closed a system as possible
4. Optimum balance between population and resources
5. Each city to confine its impacts and draw its resources from its own bioregion – see as ecosystem

Principles for city regeneration

1. Protect natural and cultural features.
2. Let topography and rural countryside define the urban form
3. Ensure development enhances environmental health
4. Intensify and diversify development
5. Maintain rural traditions
6. Work with nature
7. Education for watershed consciousness
8. Reduce car dependency

Ask students what factors describe cities. Write answers on the board.
Ask students to form groups to brainstorm remedies.

What we don't like in cities*

- Lawlessness
- Traffic jams
- Homelessness
- Freeways
- Concrete
- Drugs
- High prices
- Corruption
- Strangers
- Noise
- Large shopping malls
- Fast cars
- Beggars
- Casinos
- Skyscrapers
- Pollution
- Dangers

Over-regulation and large corporations kill cities.

What we like in cities

- Communities
- Libraries
- Anonymous
- Walk to work
- Theatres
- New people
- Parks
- Cinemas
- Ideas
- Lakes
- People size
- Street games
- Parking areas
- Street music
- Streets
- Food variety
- Street eating
- Dogs
- Good lighting
- Equal rights on
- Privacy
- Education choices
- Spirit of streets
- Opportunity and choices
- Facilities
- Lots of food trees

* Saving Our Cities from the Experts by Sam smith, in UTNE, Sept/Oct94 pages 59-82.

- Walk and ride away from traffic
- Small businesses
- Cooking smells

Urban Renewal

The key to self-regeneration of cities is self generating economies, small businesses, buying and selling in neighbourhoods. Agriculture and wilderness are essential to cities. They contribute to the ecological balance and offer solid social, economic and psychological values as well.

Oregon US requires every town to draw up and implement greenbelts and to limit or prohibit city growth. Hong Kong grew 45% of its own vegetables. New York found one quartier had 1000 pieces of spare land that could be used for gardens in Harlem.

Increase the amount of phytomass

Grow plants on cool and hot walls. Plants will deflect noise, mellow bounceback, and insulate buildings. They will take a small amount of edge off severe winds. Vegetation will collect pollution and dust. Pseudomonas in the leaves will filter positive ions and help to clean the air.

Sulphur dioxide in rain eats away mortar in many brick buildings. Vines help to preserve this structure. Develop community gardens to maximise biomass, and landscape ugly areas while giving people access to earth.

Urban forestry

Plant freeways, highways, disused power stations, railway land, schools, carparks, hospitals, nursing homes and playgrounds. Use for example, dense plantings of Casuarina and ironbark to be thinned out after four years. Sell for fuel. Save others for building timbers to grow to maturity.

Pollution resistant street trees are *Tristania conferta*, *Plantanus* spp., *Acacia longifolia*, honey locust, black locust, *Gingko biloba* (only males) *Ailanthus* and *Artemesia* spp. The black locust and gingko survived the Hiroshima Bomb.

Green guerillas in NY drill holes in concrete and throw in black locust seeds, pour boiling water over them and visit the site a few weeks later to see if the seeds have germinated and when the concrete has cracked they have a party.

Chinese tallow wood is an excellent small deciduous tree whose nuts and seeds have a wax coating from which the Chinese make wax and soap. Cold pressed seed hulls make a high grade machine oil.

Other trees include lilly pilly, chestnut, pecan and carobs. Don't use trees like mulberries that stain clothes and are squashy in streets.

Teach people to grow food to store it and sell surplus

Use garden clubs, work groups of 4-6 people, worknets for occasional help, housing co-ops and food co-ops.

Rooftop gardens are good for relaxation and stress management. Dwarf bananas, pawpaw, tomatoes, citrus-in-tubs, cucumbers, beans, peas and passionfruit can all be grown in warm climates. The roof has to be carefully designed for carrying weight and for run-off.

Small-space-gardens on windowsills and balconies can grow tomatoes, peas, annuals etc. Dwarf fruit can be kept on balconies.

Food parks Plant food trees on quiet streets. Citrus are good, so are macadamias and avocados. It is good to mix floristic trees with food producers. Pecans also make good park and street deciduous trees and they don't deprive homes of winter sun.

Retro-fit the built environment

In Berlin, Margrit Kennedy has designed and retrofitted buildings. Council has supplied money. Glass houses combined as restaurants are provided on the southern face of buildings. Solar systems are fitted. In one building waste water is completely recycled to roof gardens.

Ground level and cellars are used for housing for chickens and hares which are allowed up to courtyards.

Waste recycling of metals, glass, paper and organic matter is almost total. Berlin City Council has a huge composting centre.

'Open Space Action' – start with schools for planting and meeting in non-school hours.

Calm traffic

Use traffic calming designs. Let roads deteriorate and use fast, clean public transport. If possible use canals for transport of major goods.

Student activities

- *We need new and creative thinking about the ecology of our cities then we rethink our work, food environment and our neighbours*
- *Develop a plan to green your most detested town or city*
- *Practically now, what could you do to improve your local town or nearest city? Be specific*
- *Place a pin on a map of your town for each suburb you have visited over the past few months*
- *Make a village in your town*

UNIT 40
Last day of the PDC

Your learners are excited and tired today. You will be amazed how hard they have worked and the things they have learned that you haven't told them. It is a tidying up day for you.

The important things to cover are:

- Evaluation of the course and evaluation of their learning for themselves
- Their rights as PDC holders
- Presentation of their group work
- Future possibilities
- State of permaculture courses, locally and globally
- Final activity and party

Learning objectives

Today students have:

- A clear understanding of where they can continue their permaculture concerns and interests
- Decided where their passion is for one of the many areas of permaculture
- Prepared an item for the party which is a 'giving', a present to their colleagues and to you
- Their part of their presentation ready to support their large group presentation

Unanswered questions

Students check their notes for areas not well-covered/understood. I start the day, if I have time, with students in pairs going back through their notes to see what was not well understood or insufficient etc. This tunes them into what they have covered in the whole course before completing the evaluations. It is also steadying. They will leave the course satisfied. You don't have to answer everything just indicate where they can find information

Evaluations

I carry out four evaluations, which seems excessive. However each has a purpose.

1. Logistics for the host, whether dates, times, place, food etc. were all acceptable or better for who-ever arranged all this. I give this to the organiser.
2. Self-evaluation for students to assess their growth in knowledge over the main areas of permaculture. Students have a series of paths going up mountains. Each mountain is a main theme from the course e.g. Water. They indicate where they were at the beginning of the course in knowledge and skills, then put a second cross on the 'water' road for where they are now, at the end of the course. The distance between the two crosses is what they have learned. Students keep this.
3. Satisfaction with the course in all aspects. Students sit in two concentric circles facing each other in pairs. When they are quiet you ask the first of several questions to the inner circle, who when you say 'start', ask this question to their partner in the outer circle. The outer circle may not answer except to use body behaviour. See questions list at the end. Start with an easy question, go into the difficult question in the middle and end on a positive question. This never fails to ensure all students leave the course very content and heard.
4. Unseen teacher evaluation of student. Each students in their group presents a section of the site analysis and the final design. This is where I can see how students are suited as designers, consultants, nurserymen, overseas work and so on. It is here I can see how to frame a reference if I am asked for it in the future. Every student is suited to some role in permaculture. I have also observed and interacted with them during the course.

Where to next

The following are the traditional diploma specialist areas where it was conventional for students to spend two years in the field and document it and then apply for a diploma in up to three of these fields. They need to have a reference from their PDC teacher.

- Design consultancy
- Implementation
- Media
- Finance
- Education
- Community development

Future specialist diploma areas

Since the mid 1970s when the traditional classification was elaborated, permaculture has grown enormously and with significant specialisation in many new areas which are a complete study in themselves. These now merit a separate Diploma field and others may emerge such as Perma-occupy, which is important in the USA.

Development

- Relief – disaster – dispossessed – Intensive Care Centres
- Transition movement
- Urban permaculture design
- Rural agricultural specialisation
- Community Gardens and Schools
- GEN: the Global Ecovillage Network

Diploma pathways

Here are several options. There must be others on the other continents. These give an idea of the variety and scope of alternative diploma pathways. I believe the diversity is important. In Australia there is the formal route through the TAFE certificate system and the action learning pathway. UK has similar options.

- APT Accredited permaculture training, Received Prior Learning (RPL) through Australian TAFE system
- Permaculture Institutes
- Scandinavian model – mentoring and peer review
- Gaia University
- Chaordic Institute
- Several universities in USA e.g. See UMASS in March 2012 Permaculture UK
- Germany has designed a diploma pathway through its academies
- Austria has the Middle-Alpes Academy

Presenting final design work

- Individual design presentations – break the class into groups of about eight people for a class of say, 24 and have three groups presenting simultaneously. The time allowance for each person and group depends on how much time you have
- For large design presentations, have the class as the audience. Each group presents to everyone. I find it preferable to have all groups doing the same site. It retains interest in the topic as the observer groups see alternative designs from their own for the same piece of land. They see where the land defines the design and where they can have significant input

Final activity

Everyone form a circle. You hold a ball of coloured wool. Throw the ball to one person in the group and say what you appreciate about that person. They throw it to another person and so on. At the end you will have a network. Everyone look around the group. Then lie the wool on the ground and leave the web there until everyone has gone home. I simply say something like 'this has been an intense experience and we have lived some wonderful things as a special community which may never again meet in this way. But what we have created is never destroyed. Some may feel sad. These are just the feelings we have. Let's celebrate for ourselves and each other at our party tonight.'

FEAST and CONCERT

Party for each other and can invite family and friends. All acts are intangible gifts.

The next day or later, do not hurry students away. They seem to want to wait, chat a little and slow down and leave in a natural rhythm.

'The mind, once stretched by a new idea, never returns to its original dimensions.'

Ralph Waldo Emerson.

APPENDIX 1
Suggested course outline

Beginning Permaculture

UNIT ONE: INTRODUCTIONS
Meeting of learners and teacher. Why people have come. What they hope for. Course Outline. Timetable. Materials and references.

UNIT TWO: ECOLOGY
Permaculture is based on Ecology rather than the pure sciences. Its methods are observation and deduction rather than prescription. Its studies include – flow of energy,
cycles of matter, succession, limiting factors and perpetuation of ecosystems through design.

UNIT THREE: ETHICS, PRINCIPLES, CHARACTERISTICS
Permaculture is built on ideas and there are creative ways to use these ideas. It is concerned with clean air, water and soil, and conservation of landscapes and all species. It aims to build sustainable human societies. How to get there is suggested by characteristics, principles and ethics.

UNIT FOUR: METHODS OF DESIGN
There are several ways to design a landscape. Some of these are – observation, deduction, analysis, maps, zones and sectors.

UNIT FIVE: MAP READING
It is very useful to read maps. Map reading helps to understand ecosystems, soil types, water movement, and microclimates. It assists with water harvesting and human structures such as roads, houses and dams.

The Cultivated Ecology

UNIT SIX: WATER AND LANDSCAPE
Water is the basis of life. It is a precious mineral. Water is harvested and saved in many ways until needed by plants animals and people. Water is the basis of rehabilitating soils.

UNIT SEVEN: REJUVENATING SOILS
Soils tell you many things about plants and animals. Most soils are very damaged. There are different types of damage and different types of repair. Most soils can be improved quickly. Traditional soil classifications integrate history, use and potential.

UNIT EIGHT: DESIGNING WITH CLIMATE
Climate variation is increasing, and we need to be able to design landscapes to both avoid and/or take advantage of different types of climate. We want to reduce risk, energy use and to select appropriate plants.

UNIT NINE: DESIGN WITH MICROCLIMATES
This is where we work most closely on site. You can learn to design microclimates and to read different microclimates. A large landscape is made up of many different microclimates.

UNIT TEN: EARTHWORKS
Moving earth to buffer climate, make dams, build houses and roads can be done to increase productivity. Many mistakes can be made in earthworks and it costs a lot of money. There are some guidelines for earthworks.

UNIT ELEVEN: PLANTS IN PERMACULTURE
Plants are used for many functions in a permaculture design and are basic to every design. Propagation methods are outlined. Conservation of local and heirloom species is very important.

UNIT TWELVE: FORESTS
Understanding forests and how they work is the basis of design. A forest is an air-conditioner, soil binder, mulcher, windbreak. From knowing how forests work, landscapes are designed which are productive. Windbreaks are designed from knowledge of forests.

UNIT THIRTEEN: WINDBREAKS
Windbreaks are needed in almost every landscape. They filter the air of dust and disease. They slow down hot winds and cold winds. They protect plants, animals and buildings. Each windbreak design is site specific.

Designing Productive Landscapes

UNIT FOURTEEN: PATTERNS IN NATURE
Understanding the patterns of nature helps us to design highly productive, integrated landscapes. Patterns are linear, circles, spirals, streamlines, songs, and sayings. They all interpret landscape and improve designs.

UNIT FIFTEEN: ZONE 0 – SITING AND BUILDING HOMES
A low energy, non-polluting house can be comfortable and suit your lifestyle. A home should not be too hot or too cold and everyone can live well in it. There are principles to achieve this.

UNIT SIXTEEN: ZONE I – HOME FOOD GARDENS
Everyone, from people in the city with tiny gardens, to people with large land, can grow much of their

own food. This care for the land keeps soil and water in good condition, uses household waste, reduces food transport and, chemical use.

UNIT SEVENTEEN: ZONE II – ORCHARDS OR FOOD FORESTS
Good quality, chemically clean fruit is a security. An orchard is a food forest with many mixed species supplying fruit all year. Some non-food species are planted to provide protection and fertiliser – and later firewood.

UNIT EIGHTEEN: ZONE III – FOOD FORESTS AND SMALL ANIMALS
Poultry is best kept in an orchard to prune plants, eat pests, and provide fertiliser. They are used to 'tractor' an area, or to maintain it. Ducks, turkeys, guinea fowl and pigs are good orchard friends.

UNIT NINETEEN: ZONE II – CROPPING AND LARGE ANIMALS
There are new strategies for growing crops which conserve, build soil and reduce pests. These add to alley cropping and Fukuoka. Many crops can be grown this way and large animals are also well fed.

UNIT TWENTY: ZONE IV – RESTORATIVE FORESTS
We all use a lot of wood and other tree products in our lifetime. The structural forest is where we try to grow all our own foresty needs for bark, firewood, furniture, dyes, mulches, oils and so on. It will eventually give a very good income and improve ecosystems.

UNIT TWENTY ONE: ZONE V – PROTECT NATURAL FORESTS
These are the natural, indigenous forests of a region. They keep soil, water and animal species stable. They are usually threatened. There are sometimes remnant forests to preserve, and forests to link with wildlife corridors. There are some good ways to do this.

UNIT TWENTY TWO: BROAD CLIMATIC BIOZONES
There are many climate zones in the world. Each one has special sustainable landscapes. Soils, water use, nutrients and traditional methods have evolved over a long time and are usually sustainable. When we try to make one landscape like another, it usually fails.

Adding to Sustainability and Productivity

UNIT TWENTY THREE: SITE ANALYSIS
Designers look carefully at a site to understand its good and bad points both of which can be used in a design. There is on-site and off-site information. A site analysis is an inventory of the land from which you start to develop a design.

UNIT TWENTY FOUR: DESIGN GRAPHICS AND CREATIVE PROBLEM SOLVING
This is how to do design land and show clients what they can do to make their land more sustainable and more productive. When designing land, there are always constraints which can be hard to solve creatively. There are ways to solve problems and arrive at good solutions.

UNIT TWENTY FIVE: INCOMES FROM ACRES
Every piece of land should pay for itself. It should make a profit. This can be done without destroying the land's resources. There are many ways of making money.

UNIT TWENTY SIX: DESIGN FOR MITIGATING DISASTER
From war to drought, there are many destructive threats to human and agricultural systems. Good design strategies make landscapes strong and less likely to be damaged and more likely to recover quickly.

UNIT TWENTY SEVEN: INTEGRATED PEST MANAGEMENT
Pests are to be appreciated and managed, not eliminated. By knowing pest lifecycles, and how predators work, pests can be reduced and kept to acceptable damage levels.

UNIT TWENTY EIGHT: LIVING WITH WEEDS
Weeds are usually classified by farmers so, many useful plants are classified as weeds. Weed management means understanding the whole ecosystem. Weeds need to be managed for the benefits they bring.

UNIT TWENTY NINE: AQUACULTURE
Water systems can be highly productive. They include fish, prawns, crabs, tortoise, plants and water plants. The whole water system is an integrated ecosystem.

UNIT THIRTY: WILDLIFE FRIENDS
People and wildlife are often in conflict. Wildlife is in great danger from people. In a well-designed landscape, people and wildlife can live together.

Social Permaculture and Autonomy

UNIT THIRTY ONE: BIOREGIONS
A person cannot be self-sufficient. A region can be self-sufficient. By enriching and empowering our bioregions we can make good and strong societies.

UNIT THIRTY TWO: ECONOMICS AND ETHICAL INVESTMENT
We can use money well or badly. We can set up our financial systems to meet our own needs. We can reduce our dependence on mainstream banking.

UNIT THIRTY THREE: PERMACULTURE AT WORK
Office, shop and factory.

UNIT THIRTY FOUR: LAND OWNERSHIP
Every person has the right to use land for shelter and to meet their food needs. There are some ways of owning land which can also protect it from future misuse.

UNIT THIRTY FIVE: LAND ACCESS AND ETHICS
How can poor, indigenous and other dispossessed people get land? There are many ways. Land is a resource and not a commodity. It is there to be cared for and to meet our needs.

UNIT THIRTY SIX: LEGAL STRUCTURES
How we can protect ourselves by having good organisations and legal structures which protect our work.

UNIT THIRTY SEVEN: COMMUNITIES
Many people prefer to live in communities and these can work well or fail. This section looks at reasons why they succeed or fail.

UNIT THIRTY EIGHT: SUBURBAN PERMACULTURE
Suburban areas produce almost nothing despite having good resources in people, land and time. Suburbs can become productive parklands and good places to live.

UNIT THIRTY NINE: URBAN PERMACULTURE
There are some good Permaculture models for towns. Towns are major consumers of resources and are major polluters. They can become good places to supply many of a city's needs.

UNIT FORTY: THE LAST SESSION AND THE FUTURE

APPENDIX 2
Draft or Possible Timetable for Teaching Permaculture Design Course

DAY	SESSION I	SESSION II	SESSION III	SESSION IV
1	First class: Introductions	Ecology Principles	Ethics, Principles, Characteristics	Design Methods
2	Map Reading	Water	Water	Soils
3	Soils	Climate	Microclimates	Earthworks
4	Plants	Plants	Forests	Windbreaks
5	Windbreak	Patterns in Nature	Zone 0 Our Houses	Zone I
6	Zone I Fruit trees	Zone II Forest	Zone II Animals	Zone III Field crops and large animals
RESTDAY				
7	Zone IV Harvest Forests	Zone V Natural Forests	Site Analysis Practical	Weed Ecology
8	Wildlife Management	Integrated Pest Management	Incomes from Acres (land)	Aquaculture
9	Design for Disaster	Site analysis drawing	Practical Work on Large Design	
10	Bioregions – local wealth	Ethical Investment	Practical Work on Large Design	
11	Land Ethics & Access	Village and Commune P/C	Suburban P/C	Urban P/C
12	Design Work	Design Presentations	Evaluation	PARTY

1. Students must attend 80% of the 72 hour course to receive their Certificate.
2. Students must design: a. home garden – by individual work
b. village or commune – by group work
3. All the class groups work on the same village or commune design.
4. This Timetable is for a six hour day of theory and practical work.
5. Teachers can present this Timetable in any logical order which suits them.

Useful Resources

International Permaculture Contacts

Find your country

Worldwide Permaculture Network
www.permacultureglobal.com

Main organisations

PELUM Association (Africa)
www.pelumrd.org

Permacultura America Latina (PAL)
www.permacultura.org

Permaculture Association (Britain)
www.permaculture.org.uk

Permaculture Council for Europe
www.permaculturecouncil.eu

Permaculture in New Zealand (Inc)
www.permaculture.org.nz

Permaculture Research Institute of Australia
www.permaculturenews.org

Permaculture Tasmania Inc.
http://permaculturetas.org

Other Useful Websites

There are some incredible resources available online and we suggest you seek sources that serve your specific needs and locality in terms of climate, plants and specialist areas such as the law, for example. The following websites are examples of the kind of resources you will find with useful information and free downloadable material.

Free eBooks

www.green-shopping.co.uk/ebooks/free-ebooks.html

National Agroforestry Center
http://nac.unl.edu/index.htm

Soil and Health Library
www.soilandhealth.org

Sustainable Agriculture Research and Education
www.sare.org/Learning-Center/Multimedia

United Diversity COOP weblibrary
http://library.uniteddiversity.coop

Magazines and Journals

Permaculture Activist
www.permacultureactivist.net

Permaculture Magazine – Practical Solutions for Self-reliance
www.permaculture.co.uk

Earth Garden – Practical Solutions for Green Living
www.earthgarden.com.au

Further Reading

Most of the books listed below are available from Permanent Publication's Green Shopping website at: **www.green-shopping.co.uk** *or* **www.chelsea-green.com**

The Basics of Permaculture Design; Ross Mars; Permanent Publications, 2003.

Creating A Forest Garden – Working with nature to grow edible crops; Martin Crawford; Green Books, 2010.

Desert or Paradise – Restoring Endangered Landscapes Using Water Management, including Lake and Pond Construction; Sepp Holzer; Permanent Publications, 2012.

Designing and Maintaining Your Edible Landscape Naturally; Robert Kourik; Permanent Publications, 2004.

Earth Care Manual – A Permaculture Handbook for Britain and Other Temperate Climates; Patrick Whitefield: Permanent Publications, 2004.

Earth User's Guide to Permaculture; Rosemary Morrow; Permanent Publications, 2006.

Edible Forest Gardens – Vision and Theory v. 1: Ecological Vision and Theory for Temperate-climate Permaculture; David Jacke, Eric Toensmeier; Chelsea Green Publishing, 2006.

Edible Forest Gardens: Design and Practice v. 2: Ecological Vision and Theory for Temperate-climate Permaculture; David Jacke, Toensmeier; Chelsea Green Publishing, 2006.

Farmers of Forty Centuries –Organic Farming in China, Korea and Japan; F. H. King; Dover Publications, 2004.

Fresh Food From Small Spaces – The Square Inch Gardener's Guide to Year-Round Growing, Fermenting and Sprouting; R. J. Ruppenthal; Chelsea Green Publishing, 2007.

Gaia's Garden – A Guide to Home-scale Permaculture; Toby Hemenway; Chelsea Green Publishing, 2009.

Getting Started in Permaculture; Ross and Jenny Mars; Permanent Publications, 2007.

How to Grow Perennial Vegetables – Low-maintenance, Low-impact Vegetable Gardening; Martin Crawford, Green Books, 2012.

How to Make a Forest Garden; Patrick Whitefield; Permanent Publications, 2013.

The Living Landscape: How to Read and Understand It; Patrick Whitefield; Permanent Publications, 2013.

Paradise Lot – Two Plant Geeks, One-tenth of an Acre and the Making of an Edible Oasis in the City; Eric Toensmeier; Chelsea Green Publishing, 2013.

Perennial Vegetables; Eric Toensmeier; Chelsea Green Publishing, 2007.

People and Permaculture – Caring and Designing for Ourselves, Each Other and the Planet; Looby Macnamara; Permanent Publications, 2012.

Permaculture – A Designers' Manual; Bill Mollison; Tagari Publications; 1988.

Permaculture Design – A Step by Step Guide; Aranya; Permanent Publications, 2012.

Permaculture Garden; Graham Bell; Permanent Publications, 2004.

Permaculture In A Nutshell; Patrick Whitefield; Permanent Publications, 1993.

Permaculture In Pots – How to Grow Food in Small Urban Spaces; Juliet Kemp; Permanent Publications, 2012.

Permaculture Handbook – Garden Farming for Town and Country; New Society Publishers, 2012.

Permaculture – Principles and Pathways Beyond Sustainability; David Holmgren; Permanent Publications, 2011.

Permaculture Plants – A Selection; Jeff Nugent and Julia Boniface; Permanent Publications, 2004.

Plants For A Future – Edible and Useful Plants for a Healthy World; Ken Fern; Permanent Publications, 1997.

The Rebel Farmer; Sepp Holzer; Leopold Stocker, 2004.

Sepp Holzer's Permaculture – A Practical Guide for Farms, Orchards and Gardens; Sepp Holzer; Permanent Publications, 2010.

The Resilient Farm and Homestead – An Innovative Permaculture and Whole Systems Design Approach; Ben Falk; Chelsea Green, 2012.

Restoration Agriculture – Real-World Permaculture for Farmers; Mark Shepard; Acres USA, 2013.

Teaming with Microbes – The Organic Gardener's Guide to the Soil Food Web; Jeff Lowenfels; Timber Press, 2010.

The One-Straw Revolution; Masanobu Fukuoka; New York Review Books Classics, 2009.

The Vegetable Gardener's Guide to Permaculture – Creating an Edible Ecosystem; Christopher Shein with Julie Thompson: Timber Press, 2013.

Water For Every Farm – Yeomans Keyline Plan; P. A. Yeomans, CreateSpace Independent Publishing Platform, 2008.

The Woodland Way – a Permaculture Approach to Sustainable Woodland Management; Ben Law, Permanent Publications, 2013.

YouTube

YouTube, like all social media, is as full of useful resources as it is dross. We recommend you use your trusted networks to ask for locally relevant and quality materials on subject covered on Permaculture Design Courses. We do recommend:

Martin Crawford's Forest Garden Tour
www.youtube.com/watch?v=b_fhAch5qiY&list=TL1FIsfSrE5Bg

Paul Wheaton's channel
www.youtube.com/user/paulwheaton12

Permasolutions YouTube channel
www.youtube.com/user/Permasolutions

PRI Australia YouTube channel
www.youtube.com/user/Permasolutions

The Sepp Holzer Series on YouTube
www.youtube.com/watch?v=Bw7mQZH-fFVE&list=PL454462340A2082D0

Also please check the video content on www.permaculture.co.uk, specifically *A Farm For a Future, Greening the Desert* and Allan Savoury on Holistic Management.

Index